United States Nuclear Regulatory Commission

Protecting People and the Environment

NUREG/CR-7157
ORNL/TM-2012/96

Computational Benchmark for Estimated Reactivity Margin from Fission Products and Minor Actinides in BWR Burnup Credit

Office of Nuclear Regulatory Research

AVAILABILITY OF REFERENCE MATERIALS
IN NRC PUBLICATIONS

U.S.NRC
United States Nuclear Regulatory Commission

Protecting People and the Environment

NUREG/CR-7157
ORNL/TM-2012/96

Computational Benchmark for Estimated Reactivity Margin from Fission Products and Minor Actinides in BWR Burnup Credit

Manuscript Completed: September 2012
Date Published: February 2013

Prepared by

D. E. Mueller
J. M. Scaglione
J. C. Wagner
S. M. Bowman

Oak Ridge National Laboratory
Managed by UT-Battelle, LLC
Oak Ridge, TN 37831-6170

M. Aissa, NRC Project Manager

Prepared for
Division of Systems Analysis
Office of Nuclear Regulatory Research
U. S. Nuclear Regulatory Commission
Washington, DC 20555-0001

NRC Job Code V6061

Office of Nuclear Regulatory Research

ABSTRACT

This report proposes and documents a computational benchmark for the estimation of the additional reactivity margin available in spent nuclear fuel (SNF) from fission products and minor actinides in a burnup-credit storage/transport environment, relative to SNF compositions containing only the major actinides. The benchmark problem/configuration is a generic burnup-credit cask designed to hold 68 boiling water reactor (BWR) spent nuclear fuel assemblies. The purpose of this computational benchmark is to provide a reference configuration for the estimation of the additional reactivity margin for partial burnup credit (ISG-8) and document reference estimations of the additional reactivity margin as a function of initial enrichment, burnup, and cooling time. This benchmark model will also be used as the base case in future sensitivity studies, to be documented in a companion NUREG/CR. The geometry and material specifications are provided in sufficient detail to enable independent evaluations. Estimates of additional reactivity margin for this reference configuration may be compared to those of similar burnup-credit casks to provide an indication of the validity of design-specific estimates of fission-product margin. The reference solutions were generated with the SCALE 6.1 package. The SCALE depletion and criticality sequences have been extensively validated elsewhere for a variety of applications, including light water reactor fuel. Note that the reference solutions presented in this report are not directly or indirectly based on experimental results. Consequently, this computational benchmark cannot be used to satisfy the ANS 8.1, 8.24, and 8.27 requirements for validation of calculational methods and is not intended to be used to establish biases and bias uncertainties for burnup-credit analyses.

CONTENTS

LIST OF FIGURES

LIST OF TABLES

ACKNOWLEDGMENTS

This work was performed under contract with the Office of Nuclear Regulatory Research, U.S. Nuclear Regulatory Commission (NRC). The authors thank M. Aissa, the NRC Project Manager, D. R. Algama and R. Y. Lee of the Office of Nuclear Research (RES), T. T. Nakanishi and K. A. L. Wood of the Office of Nuclear Reactor Regulation (NRR), A. B. Barto and Z. Li of the Office of Nuclear Material Safety and Safeguards (NMSS) for their support and guidance. The authors acknowledge W. J. Marshall of ORNL for computational support. The careful reviews of the draft manuscript by W. J. Marshall and G. Radulescu are very much appreciated. Finally, the authors are thankful to A. C. Alford for her preparation of the final report.

ACRONYMS

ANS: American Nuclear Society

ANSI: American National Standards Institute

BUC: burnup credit

BWR: boiling water reactor

CRC: commercial reactor critical

DOE: U.S. Department of Energy

FP: fission product

GBC   generic burnup credit cask

Gd: gadolinium

GWd/MTU: unit of nuclear fuel burnup; gigawatt-days per initial metric ton of uranium

ISG: interim staff guidance

LWR: light water reactor

NRC: U.S. Nuclear Regulatory Commission

ORNL: Oak Ridge National Laboratory

PWR: pressurized water reactor

SNF: spent nuclear fuel

1. INTRODUCTION

This report develops and documents a computational benchmark for the estimation of the additional reactivity margin available from fission products and minor actinides, relative to calculations based on major actinides only, in a boiling water reactor (BWR) burnup-credit storage/transport environment. Relevant reference configurations for baseline and comparative analyses are provided, similar to what was developed for pressurized water reactor (PWR) fuel in NUREG/CR-6747 [1]. Herein, the major actinides are consistent with those specified in NUREG/CR-6747 (i.e., ^{234}U, ^{235}U, ^{238}U, ^{238}Pu, ^{239}Pu, ^{240}Pu, ^{241}Pu, ^{242}Pu, and ^{241}Am). Although additional reactivity margin is primarily due to fission products, a few minor actinides (^{236}U, ^{237}Np, and ^{243}Am) that have been identified as being relevant to burnup credit [2] have been included in this benchmark to provide a more complete assessment of the additional reactivity margin beyond the major actinides. The proposed benchmark problem was developed to be similar to current burnup-credit-style casks, including similar materials and dimensions. While preserving all of the important features, it approximates (or eliminates) nonessential details and proprietary information. The documentation of this computational benchmark includes all of the necessary geometric and material specifications to permit independent evaluations and sufficiently detailed reference solutions to enable meaningful comparisons. Select isotopic compositions are provided to facilitate comparisons. The reference solutions were generated with depletion (TRITON and ORIGEN-ARP) and criticality (STARBUCS and CSAS5) sequences of the SCALE 6.1 package [3] and the SCALE 238-group library based on ENDF/B-VII.0 nuclear data. It is important that the reader and potential users of this report understand that this is a *computational* benchmark, and as such, the reference solutions are based on calculations. Although the CSAS5 and TRITON sequences have been validated using laboratory critical experiments, commercial reactor criticals (CRCs), measured radiochemical assay data, and reactivity worth measurements with individual fission products important to burnup credit, the reference solutions provided herein are not directly or indirectly based on experimental results.

1.1 PURPOSE

The purpose of this computational benchmark is to provide a reference configuration to help establish a method for the estimation of reactivity margin available from fission products and minor actinides, and document estimates of the additional reactivity margin as a function of initial enrichment, burnup, and cooling time. Estimates of the additional reactivity margin for this reference configuration may be compared to those of similar burnup-credit-style casks to provide an indication of the validity of design-specific estimates of the additional reactivity margin. Detailed geometry and material specifications are provided to enable independent estimations and comparisons. As reference solutions are provided in terms of differences in effective neutron multiplication factors (Δk values), benchmarking of depletion and criticality codes, individually, is not the intent. Comparison of calculated results to the reference solutions does not satisfy the ANSI/ANS 8.1 (Section 4.3 of Ref. 4), ANSI/ANS 8.24 (Section 6 of Ref. 5), and ANSI/ANS-8.27 (Section 5 of Ref. 6) requirements for validation of calculational methods. In particular, footnote 11 of ANSI/ANS 8.1 states that validation of a calculational method by comparing the results with those of another calculational method is unacceptable. Consequently, this computational benchmark is not intended to be used to establish biases for burnup-credit analyses.

The computational benchmark model described in this report is presented for use as a reference model for sensitivity study calculations to quantify potential impacts on the calculated system k_{eff} values. Additional sensitivity studies, to be documented in a companion report, will explore the effects of varying reactor depletion parameters and burnup-credit analysis modeling approximations.

1.2 BACKGROUND

In the past, some criticality safety analyses for commercial light-water reactor (LWR) spent fuel storage and transport canisters [7,8] have assumed the SNF to be fresh (unirradiated) fuel with uniform isotopic compositions corresponding to the maximum allowable enrichment. This "*fresh-fuel assumption*" generally provides a well-defined, bounding approach for the criticality safety analysis that eliminates concerns related to the fuel operating history, and thus considerably simplifies the safety analysis. However, because this assumption ignores the decrease in fuel reactivity due to irradiation, it is very conservative and can limit the SNF capacity for a given package volume.

The concept of taking credit for the reduction in reactivity due to fuel burnup is commonly referred to as *burnup credit*. The reduction in reactivity due to fuel burnup is caused by the change in concentration (net reduction) of fissile nuclides and the production of actinide and fission-product neutron absorbers. If criticality calculations are performed based on all fissile nuclides and a limited subset of absorber nuclides present in the burned fuel, the calculated neutron multiplication factor (k_{eff}) is less than the calculated k_{eff} value using the fresh fuel assumption but remains conservative (i.e., k_{eff} is overestimated), because some absorber nuclides are omitted. The typical approach for burnup credit in storage and transportation casks has been to qualify calculated isotopic predictions via validation against destructive assay measurements from SNF samples. Thus, utilization of nuclides in a safety analysis process has been primarily limited by the availability of measured assay data. An additional consideration has been the chemical characteristics (e.g., volatility) that could potentially allow the nuclide to escape the fuel region.

Numerous studies have been performed nationally and internationally to develop a detailed understanding of the issues and phenomena associated with PWR burnup credit. The findings from these studies have supported the allowance of burnup credit for PWR fuel in transportation and storage casks. Although some studies have been performed [9, 10, 11] to support burnup credit for BWR fuel, the issues and phenomena have not been addressed in a thorough and systematic manner for relevant dry cask storage and transport systems as has been done for PWR fuel. This has been due in part to the complexity in modeling and analysis of BWR SNF caused by radial and axial variations in fuel enrichment, extensive use of fuel rods that also contain Gd_2O_3 as a burnable absorber, and significant axial moderator density variation due to a combination of two-phase flow and varying core flow. However, current conditions have caused the industry to consider burnup-credit applications for BWR SNF. This report is part of a larger effort to establish the requisite technical basis for allowance of burnup credit for BWR fuel in storage and transportation casks. The computational benchmark defined in this report provides a framework for future studies into the issues associated with implementation of burnup credit for BWR fuel.

1.3 REGULATORY GUIDANCE

The NRC issued Interim Staff Guidance 8 (ISG-8), Revision 0 [12] "Limited Burnup Credit" in May 1999, providing the first allowance of burnup credit for PWR fuel. Based on technical work performed at ORNL and elsewhere, ISG-8 has undergone two revisions, which have eliminated or lessened a number of the restrictions present in ISG-8, Revision 0. However, ISG-8 is specific to PWR fuel in transportation and storage casks, and no such similar guidance permitting burnup credit for BWR fuel in storage and transport casks has been developed. The regulatory standard review plans for dry cask storage [7] and transport [8] do not allow credit for BWR fuel burnup or fixed burnable absorbers. Relevant excerpts from NUREG-1617 include the following: "For BWR fuel assemblies, NRC staff does not currently allow any credit for burnup of the fissile material or increase in actinide or fission product poisons during irradiation; therefore, the enrichment should be that of the un-irradiated fuel" and "because of differences in net reactivity due to depletion of fissile material and burnable poisons, no credit should be taken for burnable poisons in the fuel."

2. BENCHMARK SPECIFICATION

To provide a reference burnup-credit-style cask configuration that is not constrained by unnecessary detail or proprietary information, a generic 68 BWR-assembly burnup-credit cask design was developed. This generic cask design is proposed as a reference configuration to normalize analyses and estimations of the additional reactivity margin available from fission product and minor actinide nuclides. A physical description of the generic burnup-credit (GBC) cask, referred to herein as the GBC-68 cask, is provided in this section. Reference fuel assembly dimensions, corresponding to a 10×10 BWR fuel assembly design, are also provided in this section.

2.1 GBC-68 CASK SPECIFICATION

The primary motivation for burnup credit is to increase storage and transportation cask capacities for a constant canister volume. A variety of BWR dry-storage casks are available from different vendors with significant variation in capacity (44 to 89 BWR fuel assemblies). They also vary in dimensionality and incorporate different attributes to account for design objectives. However, a common feature is the use of neutron absorber panels between fuel assembly basket cells. A basic space optimization study was conducted by Bahney and Doering [13] to evaluate assembly capacity versus amount of free volume space. Based on this assessment, 68-assembly and the 89-assembly capacity configurations had the least amount of space wasted, 24.5% and 21.6%, respectively. From a criticality perspective, having 68 assemblies or 89 assemblies represented will have little impact on the neutron leakage rate; consequently, the smaller 68-capacity configuration was selected for the generic cask design. The physical dimensions for the generic cask design are provided in Table 1 with material compositions provided in Table 2.

Table 1. Physical dimensions for the GBC-68 cask

Parameter	inches	cm
Cell inside dimension	5.9226	15.0435
Cell outside dimension	6.5132	16.5435
Cell wall (stainless steel 304) thickness	0.2953	0.75
Boral panel, overall thickness[†]	0.1010	0.2565
Boral panel, B_4C/Al thickness	0.0810	0.2057
Boral panel, Al plate thickness, 2 per panel	0.0100	0.0254
Cell pitch	6.6142	16.80
Cell height[‡]	150.0	381.0
Boral[TM] panel height	150.0	381.0
Cask inside diameter	68.8976	175.0
Cask outside diameter	84.6457	215.0
Cask wall (stainless steel 304) radial thickness	7.8740	20.0
Cask base plate thickness	11.8110	30.0
Cask lid thickness	7.8740	20.0
Cask inside height	161.7165	410.76
Active fuel height	150.0	381.0
Bottom assembly hardware length[‡‡]	5.9055	15.0
Top assembly hardware length[‡‡]	5.8110	14.76

[†] Boral[TM] is a clad composite of aluminum and boron carbide. A Boral panel or plate consists of three distinct layers. The outer layers are aluminum cladding, which form a sandwich with a central layer of uniform aggregate of fine boron carbide particles with an aluminum alloy matrix.

[‡] The cell height, Boral panel height, and active fuel height are all equivalent and their lower boundaries are coincident, 15 cm above the base plate. The 381 cm value is from the LaSalle Unit 1 reactor and was used to facilitate use of publicly available commercial reactor critical data [14] in sensitivity studies that are planned to be presented in a future report.

[‡‡] Top and bottom hardware is not modeled. This axial zone is modeled as water.

Table 2. Material compositions for the GBC-68 cask model

Isotope	Atom density (atoms/b-cm)	Weight percent
Water (Density = 0.99821 g/cm^3) [3]		
Hydrogen	6.6753×10^{-2}	11.19
Oxygen	3.3377×10^{-2}	88.81
Total	1.0013×10^{-1}	100.0
Stainless steel 304 (Density 7.94 g/cm^3) [3]		
Carbon	3.1849×10^{-4}	0.08
Silicon	1.7025×10^{-3}	1.0
Phosphorus	6.9469×10^{-5}	0.045
Chromium	1.7472×10^{-2}	19.0
Manganese	1.7407×10^{-3}	2.0
Iron	5.8546×10^{-2}	68.375
Nickel	7.7402×10^{-3}	9.5
Total	8.7589×10^{-2}	100.0
Boral panel aluminum cladding (density = 2.6972 g/cm^3)		
Aluminum (Al)	6.0200×10^{-2} [15]	100.0
Boral panel central layer (0.0200 g ^{10}B/cm^2)		
Boron-10	5.8465×10^{-3}	3.55
Boron-11	2.3532×10^{-2}	15.70
Carbon	7.3447×10^{-3}	5.35
Aluminum	4.6120×10^{-2}	75.41
Total	8.2843×10^{-2}	100.0

2.2 BWR FUEL ASSEMBLY SPECIFICATIONS

BWR fuel assembly lattices have considerable variability. BWRs have used a variety of array configurations including 6 × 6, 7 × 7, 8 × 8, 9 × 9, and 10 × 10, as well as a quad design with an internal "water cross" feature. Within each of these different lattice configurations there are other design features, including differences in numbers and placement of water rods, numbers and placement of gadolinium fuel rods, Gd_2O_3 content of gadolinium fuel rods, radial and axial variation of ^{235}U enrichments, use of part-length fuel rods, and the use of axial blankets. Each of these design features in conjunction with different reactor operating strategies will impact fuel assembly residual reactivity at discharge. Many BWR plants have a variety of lattice designs currently stored on-site, which provides an additional complexity to criticality safety evaluations for spent fuel storage and transportation.

For the GBC-68 benchmark model, a simplified BWR assembly design was adopted. The design, similar to a GE14 lattice, utilizes a 10 × 10 lattice in which eight of the fuel rods have been replaced with two large water rods. The base GBC-68 model uses the same initial ^{235}U enrichment for all fuel rods and does not include gadolinium fuel rods or part-length fuel rods or axial blankets. Evaluation of the impact of these simplifications may be performed in future studies.

Fuel assembly geometry information used in the GBC-68 benchmark model is presented in Table 3. With the exception of the active fuel length and fuel channel dimensions, fuel assembly dimensional information was taken from Table D1.A.3 in the SCALE 6.1 manual [3] for the GE14 assembly design. The active fuel length and fuel channel dimensions from the LaSalle Unit 1 CRC data [14] were used, because sensitivity studies based on the LaSalle CRC data are planned for future work. The initial (fresh) fuel assembly material composition data are provided in Table 4, where the UO_2 density used is 96% of the UO_2 theoretical maximum density (10.96 g/cm^3) [3].

Table 3. BWR fuel assembly specification

Parameter	inches	cm
Fuel pellet outside diameter	0.3449	0.876
Cladding inside diameter	0.3520	0.894
Cladding outside diameter	0.4039	1.026
Cladding thickness	0.0260	0.066
Fuel rod pitch	0.5098	1.295
Water rod inside diameter	0.9138	2.321
Water rod outside diameter	0.9925	2.521
Water rod radial thickness	0.0394	0.100
Active fuel length	150.0	381.0
Array size	10 × 10	
Number of fuel rods	92	
Number of water rods	2	
Fuel channel inner dimension	5.278	13.406
Fuel channel outer dimension	5.478	13.914
Fuel channel thickness	0.100	0.2540

Table 4. Initial material compositions for the unburned fuel assemblies

Components	Atom density (atoms/b-cm)	Weight percent
Fuel rod clad, water rod, and fuel channel (density = 6.41 g/cm^3)		
Zirconium	4.2300×10^{-2}[15]	100.0
UO$_2$, 2 wt% ^{235}U initial enrichment (density = 10.5216 g/cm^3)		
Oxygen	4.6939×10^{-2}	11.8490
^{234}U	4.2480×10^{-6}	0.0157
^{235}U	4.7527×10^{-4}	1.7630
^{236}U	2.1770×10^{-6}	0.0081
^{238}U	2.2988×10^{-2}	86.3642
Total	7.0408×10^{-2}	100.0
UO$_2$, 3 wt% ^{235}U initial enrichment (density = 10.5216 g/cm^3)		
Oxygen	4.6944×10^{-2}	11.8504
^{234}U	6.3720×10^{-6}	0.0235
^{235}U	7.1290×10^{-4}	2.6445
^{236}U	3.2654×10^{-6}	0.0122
^{238}U	2.2750×10^{-2}	85.4694
Total	7.0416×10^{-2}	100.0
UO$_2$, 4 wt% ^{235}U initial enrichment (density = 10.5216 g/cm^3)		
Oxygen	4.6950×10^{-2}	11.8517
^{234}U	8.4958×10^{-6}	0.0314
^{235}U	9.5051×10^{-4}	3.5259
^{236}U	4.3538×10^{-6}	0.0162
^{238}U	2.2512×10^{-2}	84.5748
Total	7.0425×10^{-2}	100.0
UO$_2$, 5 wt% ^{235}U initial enrichment (density = 10.5216 g/cm^3)		
Oxygen	4.6955×10^{-2}	11.8531
^{234}U	1.0620×10^{-5}	0.0392
^{235}U	1.1881×10^{-3}	4.4073
^{236}U	5.4422×10^{-6}	0.0203
^{238}U	2.2273×10^{-2}	83.6801
Total	7.0432×10^{-2}	100.0

3. ANALYSIS

3.1 COMPUTATIONAL METHODS

The computational methods necessary for this benchmark analysis include codes for fuel depletion and criticality simulation. All calculations were performed using the publicly released version of SCALE 6.1[3] and the ENDF/B-VII 238 neutron energy group library distributed with SCALE 6.1.

Depletion calculations were performed using the STARBUCS sequence and ORIGEN-ARP libraries generated using the TRITON *T-DEPL* 2-D lattice depletion sequence, which utilizes the NEWT module for flux calculations and the ORIGEN module to calculate burned fuel compositions. New ORIGEN-ARP libraries were generated for the generic 10 x 10 assembly for use in the GBC-68 benchmark model. New libraries were needed, because the GE14 ORIGEN-ARP libraries distributed with SCALE included some fuel rods with Gd_2O_3 and the existing libraries were generated without control blades being inserted. The GE14 lattice was depleted in TRITON with the control blades fully inserted for the entire depletion. To create the ORIGEN-ARP libraries used by STARBUCS, depletion calculations were performed with initial enrichments varying from 1.5 to 6 wt% ^{235}U, with moderator densities varying from 0.1 to 0.8 g/cm^3, and for fuel assembly average burnup values up to 88.5 GWd/MTU. The parameters used in the TRITON depletion model are provided in Table 5, and the fuel depletion model is shown in Figure 1. The depletion parameters used are generally consistent with the parameters used to generate the GE14 ORIGEN-ARP library distributed with SCALE 6.1 and are considered representative of average conditions fuel might experience during irradiation. These TRITON calculations produced ORIGEN-ARP libraries for initial enrichments between 1.5 and 6 wt % ^{235}U and burnup values up to 88.5 GWd/MTU. These libraries were used in STARBUCS calculations, for determining the nuclide concentrations in the burned fuel used in the GBC-68 model. Assembly average burnup values up to 60 GWd/MTU were used for the actinide and fission product reactivity margin calculations. The maximum assembly average burnup value for the non-uniform axial burnup case described in Section 3.3.3 was 79 GWd/MTU.

Table 5. Fuel depletion parameters

Parameter	Value
Pellet average fuel temperature (K)	840
Pellet average power density (MW/MTU)	40
Clad average temperature (K)	567
Moderator temperature (K)	512
Moderator density (g/cm^3)	
Water around fuel rods	0.1 to 0.8
Water inside water rod	0.776
Water between channels	0.776

A 512 K moderator temperature was used during the depletion calculations that generated the new ORIGEN-ARP libraries to be consistent with the conditions used to generate the GE14 ORIGEN-ARP libraries distributed with SCALE 6.1. This moderator temperature is low for a BWR operating around 1030 psia where T_{sat} is about 560 K. However, the moderator temperature has only a minor impact on the depletion calculations because the water density, which has the primary impact on the flux spectrum, is modeled explicitly and independent of the moderator temperature. The moderator temperature variation has only a minor impact through temperature-dependent adjustments to the scattering cross sections.

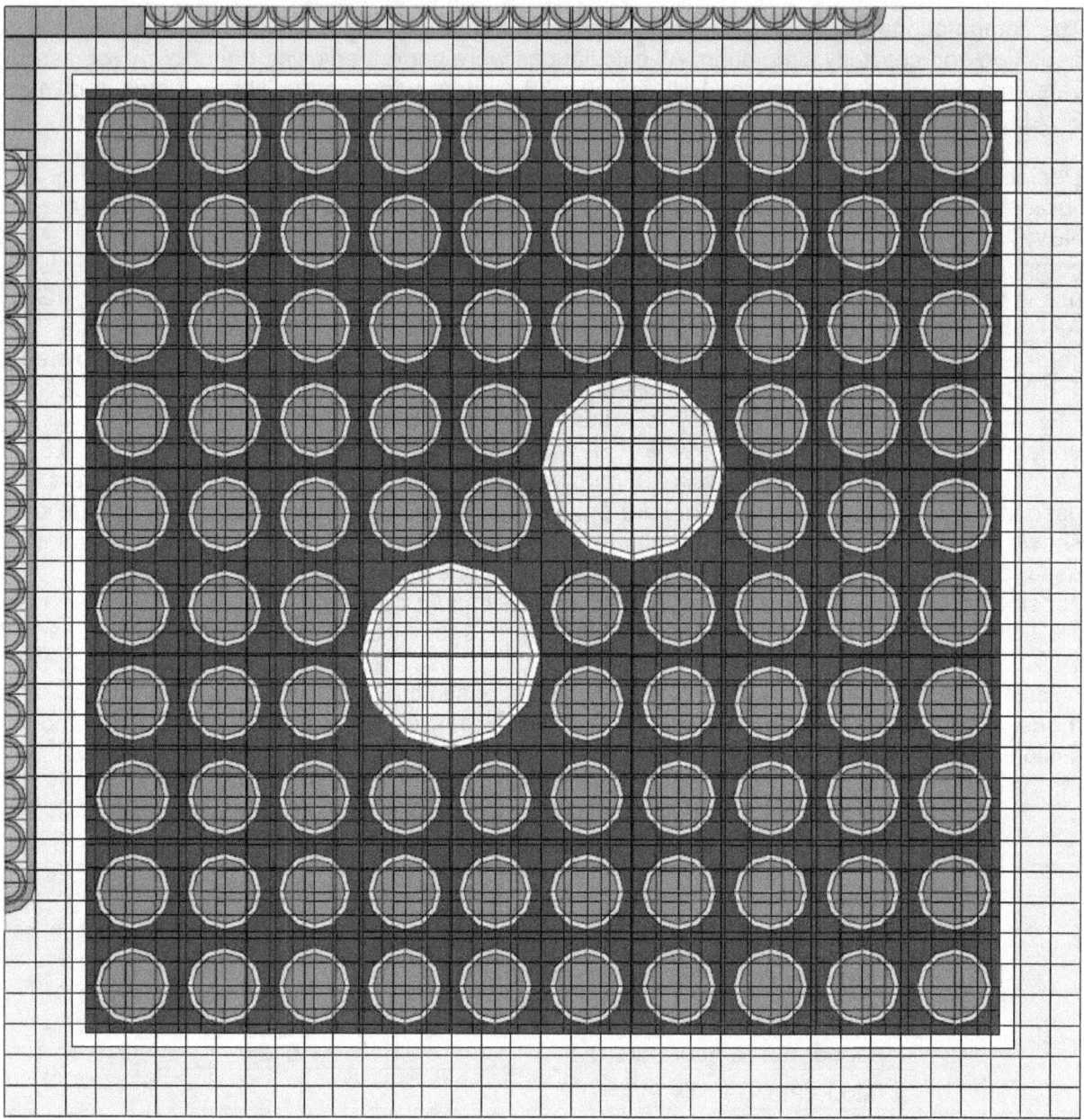

Figure 1. TRITON depletion model used to generate the rodded ORIGEN-ARP libraries.

The STARBUCS sequence automates calculation of burned fuel compositions using the ORIGEN-ARP methodology to perform a series of cross-section preparation and depletion calculations to generate a comprehensive set of spent fuel isotopic inventories for each burnup region of an assembly. The spent fuel nuclide concentrations are subsequently input to CSAS5, which performs a KENO V.a criticality calculation of the GBC-68 model to determine the neutron multiplication factor (k_{eff}) for the system. The user can specify the assembly-average irradiation history, the axial variation of the moderator density, axial and horizontal-burnup profiles, and the nuclides that are to be applied in the criticality safety analysis. Nuclide correction factors may also be applied to the predicted concentrations to account for known bias and/or uncertainty in the predicted SNF compositions. Nuclide correction factors and horizontal burnup gradients were not used for the GBC-68 calculations.

The STARBUCS sequence was used when it was necessary to calculate burned fuel compositions. The CSAS5 sequence was used when burned fuel compositions were already available from a previously performed STARBUCS calculation. All criticality calculations performed to determine k_{eff} used 11,000 neutrons per generation and skipped the first 100 generations. Calculations were run until the standard deviation on the average k_{eff} value reached 0.0001 Δk. This criterion resulted in 20 to 80 million individual neutron histories for each criticality calculation, with an average of 44 million neutron histories per KENO V.a calculation.

Example input files for the TRITON, STARBUCS, and CSAS5 calculations are provided in Appendices A, B, and C, respectively.

3.2 GBC-68 COMPUTATIONAL MODEL

Based on the benchmark specification provided in Section 2, a computational model of the GBC-68 cask, loaded with BWR fuel assemblies, was developed for KENO V.a. Cross-sectional views of the computational model are shown in Figure 2 and Figure 3. A 3-D cutaway view, generated by KENO3D, is shown in Figure 4. To aid users of this proposed benchmark in verification of their criticality models, calculated k_{eff} values corresponding to fresh fuel without axial or radial enrichment variation and without gadolinium fuel rods are provided in the next section for initial enrichments of 2, 3, 4 and 5 wt% [235]U.

The active fuel length of the assemblies is divided into 25 equal-length axial regions to facilitate the variation in axial composition due to the axial burnup distribution, which will be considered later in this report. The reference calculations were performed with a uniform axial burnup profile. Calculations using more realistic burnup profiles were also performed and are discussed later in this report.

For the criticality calculations, it is necessary to define the isotopes considered. Many studies have been performed to rank the reactivity worth of the actinide and fission product nuclides. Based on these analyses, Ref. 2 lists "prime candidates" for inclusion in burnup-credit analyses related to dry storage and transport, including several nuclides for which measured chemical assay data are not currently available in the United States. Cross-section data are generally available for all nuclides identified in Ref. 2 as being the most important for burnup-credit criticality calculations. Therefore, in this benchmark, all of the actinide and fission product nuclides identified in Ref. 2, including those for which little or no chemical assay data are available, are used in the estimation of the additional reactivity margin.

The two "nuclide sets" used here for the estimation of the additional reactivity margin are listed in Table 6. The first set, which corresponds to the major actinides specified in a DOE topical report [16], is used for the reference actinide-only calculations. The second set includes all of the actinide and fission product nuclides identified in Ref. 2 as being important for burnup credit criticality calculations; the first set is a subset of the second set. These nuclide sets are consistent with those used in NUREG/CR-6747[1], NUREG/CR-7108[17], and NUREG/CR-7109[18].

For the purpose of this benchmark report and the cited results, the additional reactivity margin available from fission products and minor actinides is due to the nuclides that are exclusive to the second set. These "additional nuclides," which are exclusive to the second set, are listed in Table 7 for clarity and are designated as "set 3." Throughout this report, where reference is made to the additional reactivity margin due to the additional actinide and fission product nuclides, the additional reactivity margin is due to the nuclides listed in Table 6.

Finally, it should be noted that these "nuclide sets" are defined for the purposes of this analysis only; other terminology and specific sets of nuclides have been defined and used by individuals studying burnup-credit phenomena.

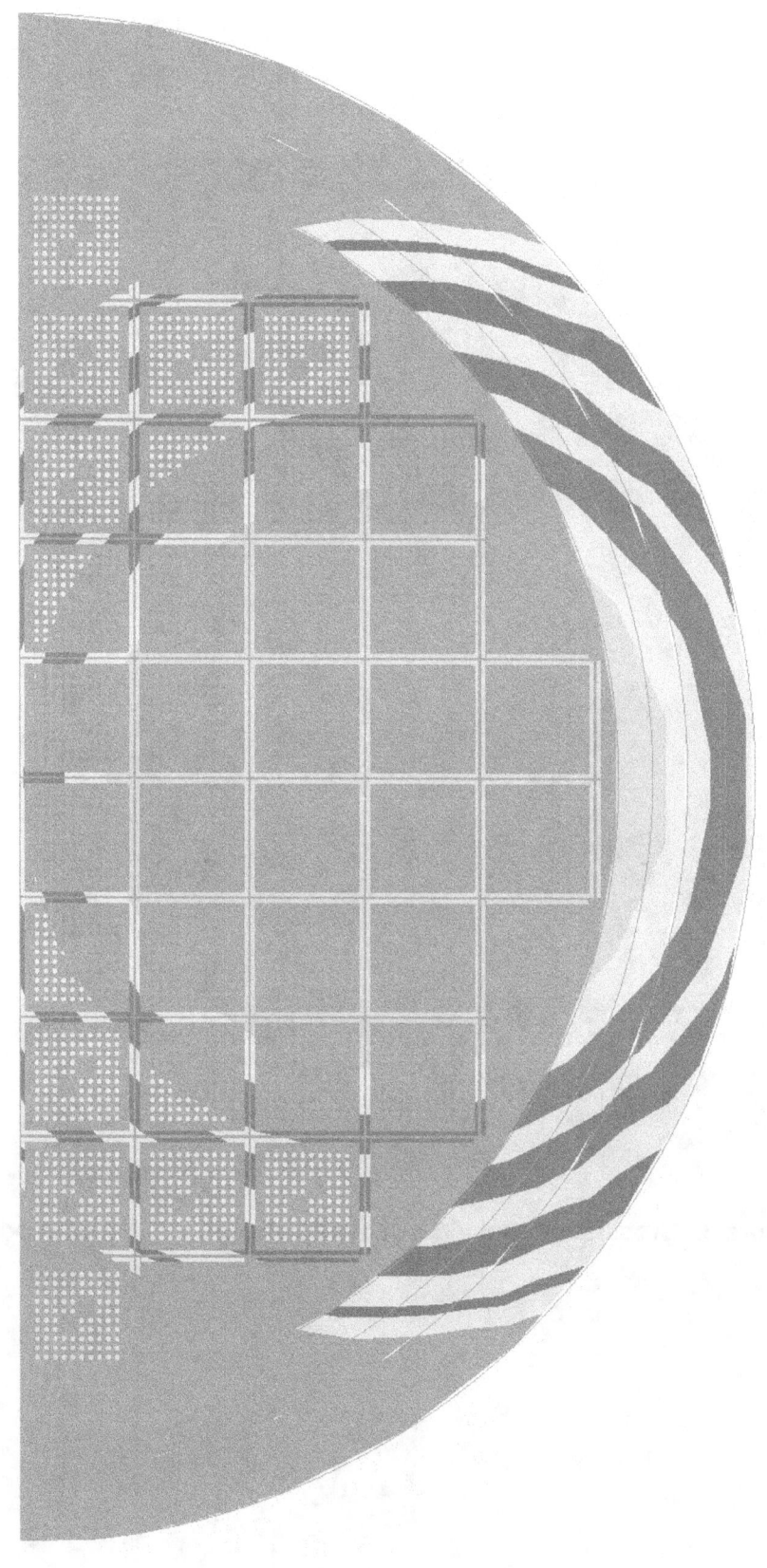

Figure 2. Radial cross section of the GBC-68 model with reflective left boundary condition.

Figure 3. Cross-sectional view of assembly cell in GBC-68 model.

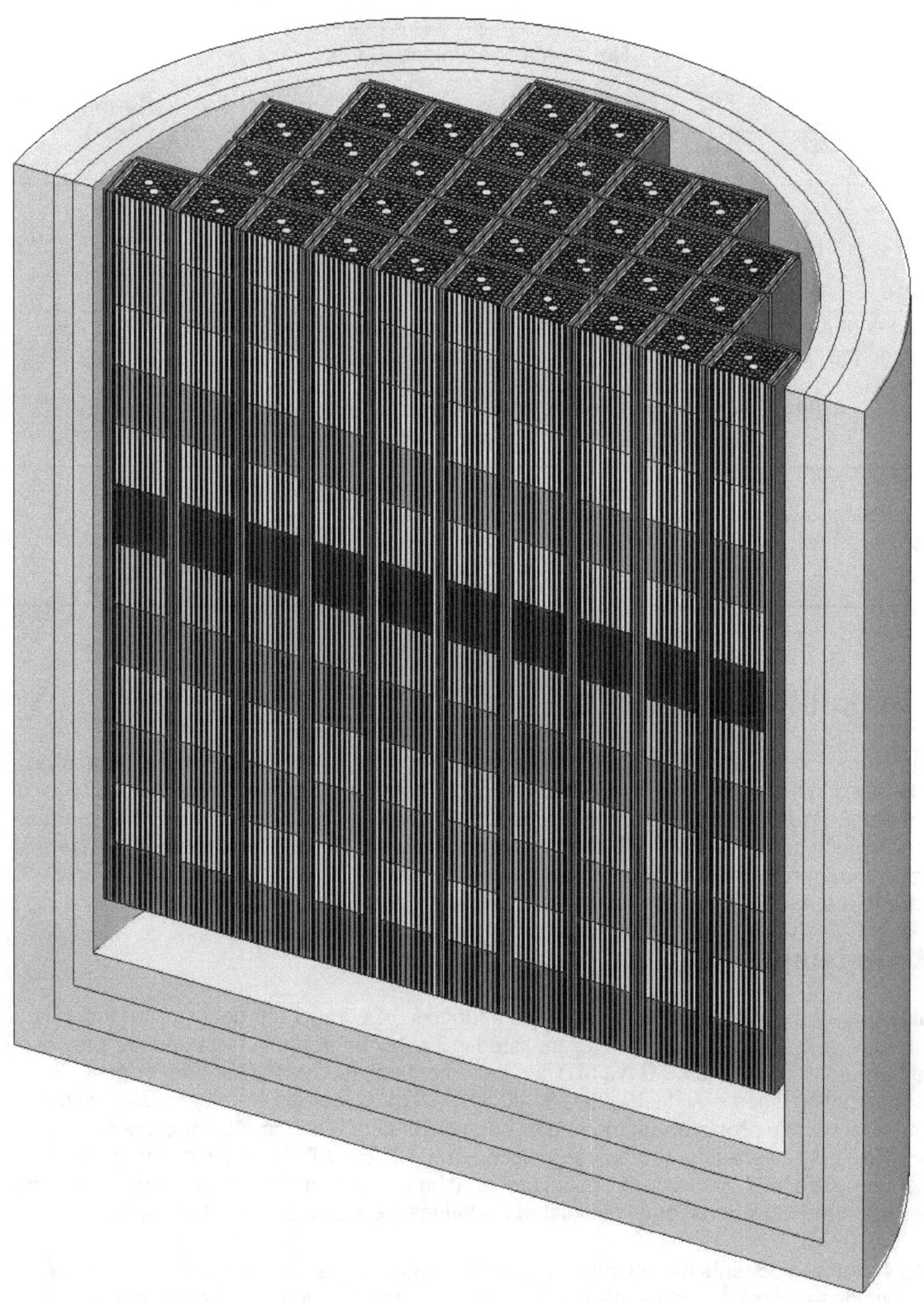

Figure 4. Cutaway view of bottom part GBC-68 model. Poison panel and cell walls removed to reveal fuel assembly axial zoning. Note that the reference cases utilized a single axial zone.

Table 6. Nuclide sets defined for the benchmark problem analysis

Set 1: Major actinides (9 total)									
^{234}U	^{235}U	^{238}U	^{238}Pu	^{239}Pu	^{240}Pu	^{241}Pu	^{242}Pu	^{241}Am	O^{+}
Set 2: Actinides and major fission products (28 total)									
^{234}U	^{235}U	^{236}U	^{238}U	^{238}Pu	^{239}Pu	^{240}Pu	^{241}Pu	^{242}Pu	^{241}Am
^{243}Am	^{237}Np	^{95}Mo	^{99}Tc	^{101}Ru	^{103}Rh	^{109}Ag	^{133}Cs	^{147}Sm	^{149}Sm
^{150}Sm	^{151}Sm	^{152}Sm	^{143}Nd	^{145}Nd	^{151}Eu	^{153}Eu	^{155}Gd	O^{+}	

$^{+}$Oxygen is neither an actinide nor a fission product, but is included in this list because it is included in the calculations.

Table 7. Nuclides in "set 3," on which the additional reactivity margin available from fission products and minor actinides is based

Set 3: Minor actinides and major fission products (19 total)									
^{236}U	^{243}Am	^{237}Np	^{95}Mo	^{99}Tc	^{101}Ru	^{103}Rh	^{109}Ag	^{133}Cs	^{147}Sm
^{149}Sm	^{150}Sm	^{151}Sm	^{152}Sm	^{143}Nd	^{145}Nd	^{151}Eu	^{153}Eu	^{155}Gd	

3.3   RESULTS

Results for the computational benchmark are presented in this section. Based on the nuclide sets identified in the previous section, calculated k_{eff} values are provided as a function of initial enrichment, burnup, and cooling times, within the ranges relevant to storage and transportation. A rather large volume of results is included for completeness. It is not anticipated that users of this benchmark problem will attempt to reproduce the complete set of results, but rather compare to a subset of the reference results that are relevant to their application.

3.3.1   Reference Results

Calculated k_{eff} values for the GBC-68 cask as a function of burnup and cooling time for initial enrichments of 2, 3, 4, and 5 wt % ^{235}U are listed in Tables 8 through 11. Values are provided for the burnup range of 0 to 60 GWd/MTU, in increments of 10 GWd/MTU, and for post-irradiation cooling times of 0, 5, 10, 20, and 40 years. The k_{eff} values were calculated with a uniform axial burnup profile and a moderator density value of 0.6 g/cm^{3}. Control blades were present (i.e., fully inserted for the entire depletion) in ORIGEN-ARP calculations to generate nuclide concentrations based on the initial fuel enrichment, assembly burnup, and cooling time values. Standard deviations for all calculated k_{eff} values were approximately 0.00010.

The individual components of reduction in k_{eff} (Δk), relative to fresh fuel, associated with (a) the major actinides and (b) the additional nuclides as a function of burnup and cooling time are listed in Tables 12 through 15 for each initial fuel enrichment. The second column from the left in Tables 12 through 15 lists the Δk reductions due to the presence of the major actinides alone (nuclide set 1, see Table 6), while the third column lists the Δk reduction due to the presence of the fission products and additional actinides (i.e., due to the nuclides present in set 3, see Table 7). Thus, the results listed in the third column may be interpreted as the additional Δk margin associated with the fission products and additional actinides. The fourth column from the left in Table 12 through Table 15 lists the total Δk reduction as a function of burnup for the cooling times considered. Finally, the two columns on the right-hand side of the table list the percent contributions from the two sets of nuclides to the total Δk reduction, and thus provide an assessment of the reactivity reduction, relative to fresh fuel, associated with (a) the major actinides and (b) the fission products and additional nuclides.

When associating practical meaning to these results, it is important that the reader and potential users of this report understand that this is a **computational** benchmark, and as such, the reference solutions are based on calculations alone (e.g., no isotopic correction factors are applied). The reference solutions are not directly or indirectly based on experimental results. However, note that the computational tools used to generate the reference solutions have been validated elsewhere [18,19].

As noted above, the reference results presented in Tables 8 through 15 are for a uniform axial burnup profile. Because conservative axial burnup profiles are not available, the reference results were generated using a uniform axial profile. Future BWR burnup-credit work should include identification of conservative axial burnup profiles and generation of reference results using those axial burnup profiles.

In addition to the uniform axial burnup distribution simplification, the computational benchmark uses a uniform axial moderator density for simplification. This is necessary because, as with axial burnup distribution data, very little work has been done to collect and evaluate realistic axial moderator density profiles. The potential impact of varying the average moderator density and of modeling the axial distribution of moderator density will be addressed in future reports designed to improve the bases for BWR burnup credit.

Table 8. k_{eff} values for the GBC-68 cask as a function of burnup and cooling time for 2 wt% ^{235}U initial enrichment

Major actinides (nuclide set 1, see Table 6)					
Cooling time (years)	0	5	10	20	40
Burnup (GWd/MTU)	k_{eff} values*				
0	0.77303	0.77303	0.77303	0.77303	0.77303
10	0.73639	0.73359	0.72996	0.72497	0.71992
20	0.69217	0.68232	0.67262	0.65872	0.64533
30	0.65632	0.64069	0.62630	0.60585	0.58546
40	0.63322	0.61353	0.59593	0.57068	0.54558
50	0.62082	0.59861	0.57869	0.55013	0.52179
60	0.61586	0.59202	0.57074	0.54020	0.50961
Actinides and major fission products (nuclide set 2, see Table 6)					
0	0.77303	0.77303	0.77303	0.77303	0.77303
10	0.71181	0.70289	0.69840	0.69294	0.68785
20	0.65391	0.63560	0.62443	0.60999	0.59692
30	0.60817	0.58227	0.56553	0.54450	0.52542
40	0.57756	0.54625	0.52593	0.50049	0.47721
50	0.55906	0.52400	0.50148	0.47350	0.44795
60	0.54892	0.51158	0.48744	0.45796	0.43106

*Standard deviation is approximately 0.0001 for all calculations

Table 9. k_{eff} **values for the GB C-68 cask as a function of burnup and cooling time for 3 wt% ^{235}U initial enrichment**

Major actinides (nuclide set 1, see Table 6)					
Cooling time (years)	0	5	10	20	40
Burnup (GWd/MTU)	k_{eff} values*				
0	0.87624	0.87624	0.87624	0.87624	0.87624
10	0.83454	0.83328	0.83126	0.82834	0.82561
20	0.78600	0.77877	0.77195	0.76254	0.75363
30	0.73777	0.72498	0.71356	0.69780	0.68231
40	0.69550	0.67821	0.66285	0.64136	0.62023
50	0.66306	0.64214	0.62399	0.59806	0.57272
60	0.64214	0.61908	0.59889	0.56996	0.54121
Actinides and major fission products (nuclide set 2, see Table 6)					
0	0.87624	0.87624	0.87624	0.87624	0.87624
10	0.80845	0.80168	0.79866	0.79567	0.79257
20	0.74503	0.73009	0.72216	0.71203	0.70338
30	0.68519	0.66292	0.64963	0.63293	0.61818
40	0.63456	0.60594	0.58834	0.56607	0.54628
50	0.59618	0.56272	0.54182	0.51586	0.49255
60	0.57056	0.53422	0.51108	0.48283	0.45724

*Standard deviation is approximately 0.0001 for all calculations

Table 10. k_{eff} values for the GBC-68 cask as a function of burnup and cooling time for 4 wt% ^{235}U initial enrichment

Major actinides (nuclide set 1, see Table 6)					
Cooling time (years)	0	5	10	20	40
Burnup (GWd/MTU)	k_{eff} values*				
0	0.94283	0.94283	0.94283	0.94283	0.94283
10	0.90466	0.90380	0.90241	0.90076	0.89892
20	0.86015	0.85491	0.85023	0.84382	0.83729
30	0.81393	0.80387	0.79532	0.78323	0.77161
40	0.76734	0.75282	0.74028	0.72273	0.70570
50	0.72381	0.70533	0.68951	0.66723	0.64547
60	0.68789	0.66636	0.64755	0.62129	0.59546
Actinides and major fission products (nuclide set 2, see Table 6)					
0	0.94283	0.94283	0.94283	0.94283	0.94283
10	0.87755	0.87177	0.86989	0.86774	0.86614
20	0.81780	0.80611	0.80031	0.79316	0.78722
30	0.75899	0.74066	0.73015	0.71734	0.70591
40	0.70263	0.67746	0.66269	0.64438	0.62793
50	0.65165	0.62134	0.60230	0.57965	0.55930
60	0.61060	0.57592	0.55452	0.52822	0.50466

*Standard deviation is approximately 0.0001 for all calculations

Table 11. k_{eff} values for the GBC-68 cask as a function of burnup and cooling time for 5 wt% ^{235}U initial enrichment

Major actinides (nuclide set 1, see Table 6)					
Cooling time (years)	0	5	10	20	40
Burnup (GWd/MTU)	k_{eff} values*				
0	0.99048	0.99048	0.99048	0.99048	0.99048
10	0.95634	0.95594	0.95486	0.95384	0.95284
20	0.91693	0.91296	0.90967	0.90472	0.90017
30	0.87603	0.86791	0.86112	0.85196	0.84290
40	0.83235	0.82040	0.81035	0.79646	0.78269
50	0.78825	0.77202	0.75854	0.74015	0.72182
60	0.74533	0.72573	0.70932	0.68665	0.66397
Actinides and major fission products (nuclide set 2, see Table 6)					
0	0.99048	0.99048	0.99048	0.99048	0.99048
10	0.92921	0.92403	0.92260	0.92110	0.92015
20	0.87425	0.86466	0.85996	0.85476	0.85022
30	0.82004	0.80458	0.79646	0.78634	0.77752
40	0.76554	0.74412	0.73199	0.71679	0.70370
50	0.71246	0.68544	0.66915	0.64965	0.63246
60	0.66326	0.63115	0.61182	0.58828	0.56737

*Standard deviation is approximately 0.0001 for all calculations

Table 12. Individual components of the reduction in k_{eff} as a function of burnup and cooling time for fuel of 2 wt% ^{235}U initial enrichment

Burnup (GWd/MTU)	Δk values due to the various nuclide sets			Contribution (%) to total reduction in k_{eff}	
	Major actinides (set 1)	Additional nuclides (set 3)	Total (set 2)	Major actinides (set 1)	Additional nuclides (set 3)
0-year cooling time					
10	0.03664	0.02458	0.06122	59.84	40.16
20	0.08085	0.03827	0.11912	67.88	32.12
30	0.11670	0.04815	0.16485	70.79	29.21
40	0.13980	0.05567	0.19547	71.52	28.48
50	0.15221	0.06175	0.21396	71.14	28.86
60	0.15717	0.06694	0.22411	70.13	29.87
5-year cooling time					
10	0.03943	0.03070	0.07013	56.23	43.77
20	0.09071	0.04672	0.13743	66.00	34.00
30	0.13234	0.05842	0.19076	69.37	30.63
40	0.15950	0.06728	0.22678	70.33	29.67
50	0.17442	0.07461	0.24903	70.04	29.96
60	0.18101	0.08043	0.26144	69.23	30.77
10-year cooling time					
10	0.04306	0.03156	0.07462	57.71	42.29
20	0.10040	0.04819	0.14859	67.57	32.43
30	0.14673	0.06076	0.20750	70.72	29.28
40	0.17710	0.07000	0.24710	71.67	28.33
50	0.19434	0.07721	0.27155	71.57	28.43
60	0.20229	0.08330	0.28559	70.83	29.17
20-year cooling time					
10	0.04806	0.03203	0.08009	60.01	39.99
20	0.11431	0.04872	0.16303	70.11	29.89
30	0.16717	0.06135	0.22853	73.15	26.85
40	0.20235	0.07018	0.27253	74.25	25.75
50	0.22290	0.07662	0.29952	74.42	25.58
60	0.23283	0.08224	0.31507	73.90	26.10
40-year cooling time					
10	0.05310	0.03208	0.08518	62.34	37.66
20	0.12769	0.04841	0.17611	72.51	27.49
30	0.18757	0.06004	0.24761	75.75	24.25
40	0.22744	0.06837	0.29581	76.89	23.11
50	0.25124	0.07383	0.32508	77.29	22.71
60	0.26342	0.07855	0.34197	77.03	22.97

Table 13. Individual components of the reduction in k_{eff} as a function of burnup and cooling time for fuel of 3 wt% ^{235}U initial enrichment

| Burnup (GWd/MTU) | Δk values due to the various nuclide sets | | | Contribution (%) to total reduction in k_{eff} | |
	Major actinides (set 1)	Additional nuclides (set 3)	Total (set 2)	Major actinides (set 1)	Additional nuclides (set 3)
0-year cooling time					
10	0.04170	0.02609	0.06779	61.51	38.49
20	0.09023	0.04098	0.13121	68.77	31.23
30	0.13847	0.05257	0.19104	72.48	27.52
40	0.18074	0.06094	0.24168	74.79	25.21
50	0.21318	0.06688	0.28006	76.12	23.88
60	0.23409	0.07158	0.30568	76.58	23.42
5-year cooling time					
10	0.04296	0.03159	0.07455	57.62	42.38
20	0.09746	0.04868	0.14614	66.69	33.31
30	0.15125	0.06207	0.21332	70.90	29.10
40	0.19803	0.07227	0.27030	73.26	26.74
50	0.23410	0.07942	0.31352	74.67	25.33
60	0.25716	0.08485	0.34201	75.19	24.81
10-year cooling time					
10	0.04497	0.03260	0.07757	57.98	42.02
20	0.10428	0.04980	0.15408	67.68	32.32
30	0.16268	0.06393	0.22661	71.79	28.21
40	0.21339	0.07451	0.28790	74.12	25.88
50	0.25225	0.08216	0.33441	75.43	24.57
60	0.27734	0.08781	0.36515	75.95	24.05
20-year cooling time					
10	0.04790	0.03267	0.08057	59.45	40.55
20	0.11369	0.05052	0.16421	69.24	30.76
30	0.17844	0.06487	0.24331	73.34	26.66
40	0.23488	0.07529	0.31017	75.73	24.27
50	0.27818	0.08219	0.36037	77.19	22.81
60	0.30628	0.08712	0.39340	77.85	22.15
40-year cooling time					
10	0.05062	0.03304	0.08366	60.51	39.49
20	0.12261	0.05025	0.17285	70.93	29.07
30	0.19392	0.06413	0.25806	75.15	24.85
40	0.25600	0.07396	0.32996	77.59	22.41
50	0.30352	0.08017	0.38368	79.11	20.89
60	0.33503	0.08396	0.41899	79.96	20.04

Table 14. Individual components of the reduction in k_{eff} as a function of burnup and cooling time for fuel in 4 wt% ^{235}U initial enrichment

Burnup (GWd/MTU)	Δk values due to the various nuclide sets			Contribution (%) to total reduction in k_{eff}	
	Major actinides (set 1)	Additional nuclides (set 3)	Total (set 2)	Major actinides (set 1)	Additional nuclides (set 3)
0-year cooling time					
10	0.03817	0.02711	0.06527	58.48	41.52
20	0.08268	0.04235	0.12503	66.12	33.88
30	0.12890	0.05494	0.18384	70.11	29.89
40	0.17549	0.06471	0.24020	73.06	26.94
50	0.21902	0.07216	0.29118	75.22	24.78
60	0.25494	0.07729	0.33223	76.74	23.26
5-year cooling time					
10	0.03903	0.03203	0.07106	54.93	45.07
20	0.08792	0.04880	0.13672	64.31	35.69
30	0.13896	0.06321	0.20217	68.73	31.27
40	0.19001	0.07535	0.26537	71.60	28.40
50	0.23750	0.08399	0.32149	73.88	26.12
60	0.27647	0.09045	0.36691	75.35	24.65
10-year cooling time					
10	0.04042	0.03252	0.07294	55.42	44.58
20	0.09260	0.04992	0.14252	64.97	35.03
30	0.14751	0.06518	0.21268	69.36	30.64
40	0.20255	0.07759	0.28013	72.30	27.70
50	0.25332	0.08721	0.34053	74.39	25.61
60	0.29528	0.09302	0.38830	76.04	23.96
20-year cooling time					
10	0.04207	0.03301	0.07509	56.03	43.97
20	0.09900	0.05066	0.14966	66.15	33.85
30	0.15959	0.06590	0.22549	70.78	29.22
40	0.22010	0.07836	0.29845	73.75	26.25
50	0.27560	0.08758	0.36318	75.89	24.11
60	0.32154	0.09307	0.41461	77.55	22.45
40-year cooling time					
10	0.04391	0.03278	0.07669	57.25	42.75
20	0.10554	0.05007	0.15561	67.82	32.18
30	0.17122	0.06570	0.23692	72.27	27.73
40	0.23713	0.07777	0.31489	75.30	24.70
50	0.29736	0.08617	0.38353	77.53	22.47
60	0.34737	0.09080	0.43817	79.28	20.72

Table 15. Individual components of the reduction in k_{eff} as a function of burnup and cooling time for fuel of 5 wt% ^{235}U initial enrichment

Burnup (GWd/MTU)	Δk values due to the various nuclide sets			Contribution (%) to total reduction in k_{eff}	
	Major actinides (set 1)	Additional nuclides (set 3)	Total (set 2)	Major actinides (set 1)	Additional nuclides (set 3)
0-year cooling time					
10	0.03414	0.02713	0.06127	55.72	44.28
20	0.07355	0.04268	0.11623	63.28	36.72
30	0.11445	0.05599	0.17043	67.15	32.85
40	0.15812	0.06681	0.22494	70.30	29.70
50	0.20222	0.07579	0.27801	72.74	27.26
60	0.24515	0.08206	0.32721	74.92	25.08
5-year cooling time					
10	0.03454	0.03191	0.06645	51.97	48.03
20	0.07752	0.04830	0.12581	61.61	38.39
30	0.12256	0.06333	0.18589	65.93	34.07
40	0.17008	0.07628	0.24636	69.04	30.96
50	0.21845	0.08659	0.30504	71.61	28.39
60	0.26475	0.09458	0.35933	73.68	26.32
10-year cooling time					
10	0.03561	0.03226	0.06788	52.47	47.53
20	0.08081	0.04972	0.13052	61.91	38.09
30	0.12936	0.06466	0.19401	66.67	33.33
40	0.18012	0.07837	0.25849	69.68	30.32
50	0.23194	0.08939	0.32133	72.18	27.82
60	0.28116	0.09750	0.37865	74.25	25.75
20-year cooling time					
10	0.03663	0.03274	0.06937	52.80	47.20
20	0.08576	0.04996	0.13572	63.19	36.81
30	0.13851	0.06562	0.20414	67.85	32.15
40	0.19402	0.07967	0.27369	70.89	29.11
50	0.25032	0.09051	0.34083	73.45	26.55
60	0.30383	0.09837	0.40220	75.54	24.46
40-year cooling time					
10	0.03763	0.03269	0.07033	53.51	46.49
20	0.09031	0.04995	0.14025	64.39	35.61
30	0.14757	0.06539	0.21296	69.30	30.70
40	0.20779	0.07899	0.28678	72.46	27.54
50	0.26866	0.08936	0.35801	75.04	24.96
60	0.32651	0.09660	0.42311	77.17	22.83

3.3.2 Related Information

Additional results and supplementary discussion, which should not be considered part of this computational benchmark, are available in Appendix D. Discussion of the reference results, including graphical representations and relevant observations, is given in Appendix D.

3.3.3 Additional Results with Non-Uniform Axial Burnup

Considerable work has been done to characterize and develop axial burnup profiles for PWR fuel. A review of this subject as applied to PWRs is provided in NUREG/CR-6801 [19]. For BWRs, assessment of the impact of axial burnup distribution is significantly more complicated than for PWRs because of the extensive use of axial design features such as axially varying ^{235}U enrichment, part-length fuel rods, axial blankets of natural uranium, and axial variation of initial gadolinia burnable absorber in some fuel rods. Reactor operations affect axial burnup by the use of control blades for long-term reactivity control and the significant axial variation in effective moderator density, which is caused by both reduction in density due to increased temperature and the presence of steam voids generated by boiling. For example, in the LaSalle Unit 1 CRC data, the effective moderator density varies from 0.74 g/cm^3 at the bottom of the core to 0.17 g/cm^3 at the tops of some assemblies [14].

Due to the axially dependent complexities associated with BWR SNF and the lack of an established database of evaluated axial burnup distribution information for BWR fuel, the GBC-68 benchmark model in Sect. 3.2 did not include axial design features or axial burnup or moderator density distributions. Recognizing that these issues are still relevant for BWR SNF, two additional series of calculations were performed to provide an example of the potential impact of using axial burnup distributions and of modeling natural uranium blankets.

To demonstrate the potential effects associated with axial burnup distribution, a few calculations were performed with axial burnup distributions extracted from the CRC data for LaSalle Unit 1 While it is not clear from the reference, these data were most likely inferred using a nodal code and Transverse Incore Probe (TIP) measurements. The 25-node burnup data for assembly ID B2 from Ref. 14 were used, because this assembly reached the highest assembly average burnup in the LaSalle Unit 1 CRC data. The axial burnup distributions used in this report are shown in Figure 5 and Table 16. These five profiles had assembly average burnups closest to the assembly average burnup values of interest.

From Figure 5, there appears to be an axial shift in the power distribution between the 10 and 20 GWd/MTU profiles. This is likely due to control blade withdrawal near the assembly. Because of the extensive and long-term usage of control blades to control reactivity and power peaking, axial power shifts due to control blade movements are likely seen by most assemblies. The example provided in this section is a single sample being used to show that modeling of axial burnup distributions will be necessary for application of BWR burnup-credit.

In the first series of calculations, the burnup-dependent normalized axial burnup profiles (see Table 16) from LaSalle Unit 1 assembly B2 were applied to the benchmark model with initial enrichments of 2, 3, 4, and 5 wt % ^{235}U, burned to 10, 20, 30, 40, 50, and 60 GWd/MTU, and cooled for 5 years following irradiation. Note that these calculations did not include modeling of natural uranium blankets. The resulting k_{eff} values and the Δk values with respect to the base model are provided in the middle two columns of Tables 17 through 20. The comparison demonstrates that the "end effect" previously seen for PWR burnup credit also exists for BWR burnup credit. Note that, as was seen for PWR burnup credit, the reactivity increases are significantly larger for the actinide plus fission products cases (set 2 nuclides) than for the major actinide only cases (set 1 nuclides).

A description of LaSalle Unit 1 assembly B2 design is provided in Table 21. Note that only the axial burnup distribution data were used. No attempt was made to model the axially dependent enrichments, axial variation in gadolinium rod loading, or the axial dependence of the moderator density. Comprehensive detailed studies of the impact of such axially varying features should be performed, but is beyond the scope of the work presented in this report. Modeling the additional axial detail shown in Table 21 would not impact the general conclusion that it will be necessary to model axial burnup distributions for BWR BUC at middle to high burnup values.

The actual relative burnup in the natural uranium blanket zones is significantly lower than it would be for unblanketed fuel. An additional series of calculations was performed with natural uranium blankets modeled in the two end fuel zones. These calculations provide a more realistic assessment of the impact of axial burnup distribution on the GBC-68 models. The results are provided in the two right-most columns of Tables 17 through 20. These results from the models with blankets show an increase in k_{eff} that is significantly smaller than the increase observed when a blanketed normalized assembly burnup profile is applied to unblanketed fuel.

Note that the reactivity worth of the fission products increases with burnup and that, with the axial burnup distribution present, the lower burnup region near the top of the assembly controls the reactivity. Thus, for a given assembly-average burnup, the reactivity margin due to fission products will be overestimated if the axial burnup distribution is not included in the model. Therefore, it should be emphasized that the computational benchmark results with uniform axial burnup are provided to enable comparison with a simpler benchmark problem and should not be considered to be representative of actual reactivity margins. This modeling simplification reduces the volume of composition data by a factor of 25 (i.e., the number of axial regions used to represent the axial burnup distribution). Nuclide compositions for fuel with initial enrichment of 4 wt % ^{235}U at various burnup and cooling time combinations are provided in Appendix E to enable comparisons of calculated spent fuel compositions.

Comparison of the results within Tables 17 through 20 shows that, with the uniform axial burnup distribution, the calculated total reactivity reduction due to actinides and actinides with fission products is overestimated for burnups greater than approximately 10 GWd/MTU. Thus, for typical discharge burnups (30–50 GWd/MTU), the individual components of reactivity reduction associated with (a) the major actinides and (b) the additional nuclides are both overestimated with the uniform axial burnup distribution. Additionally, at the higher burnup values the k_{eff} values calculated for the uniform axial burnup distribution appear to be significantly non-conservative.

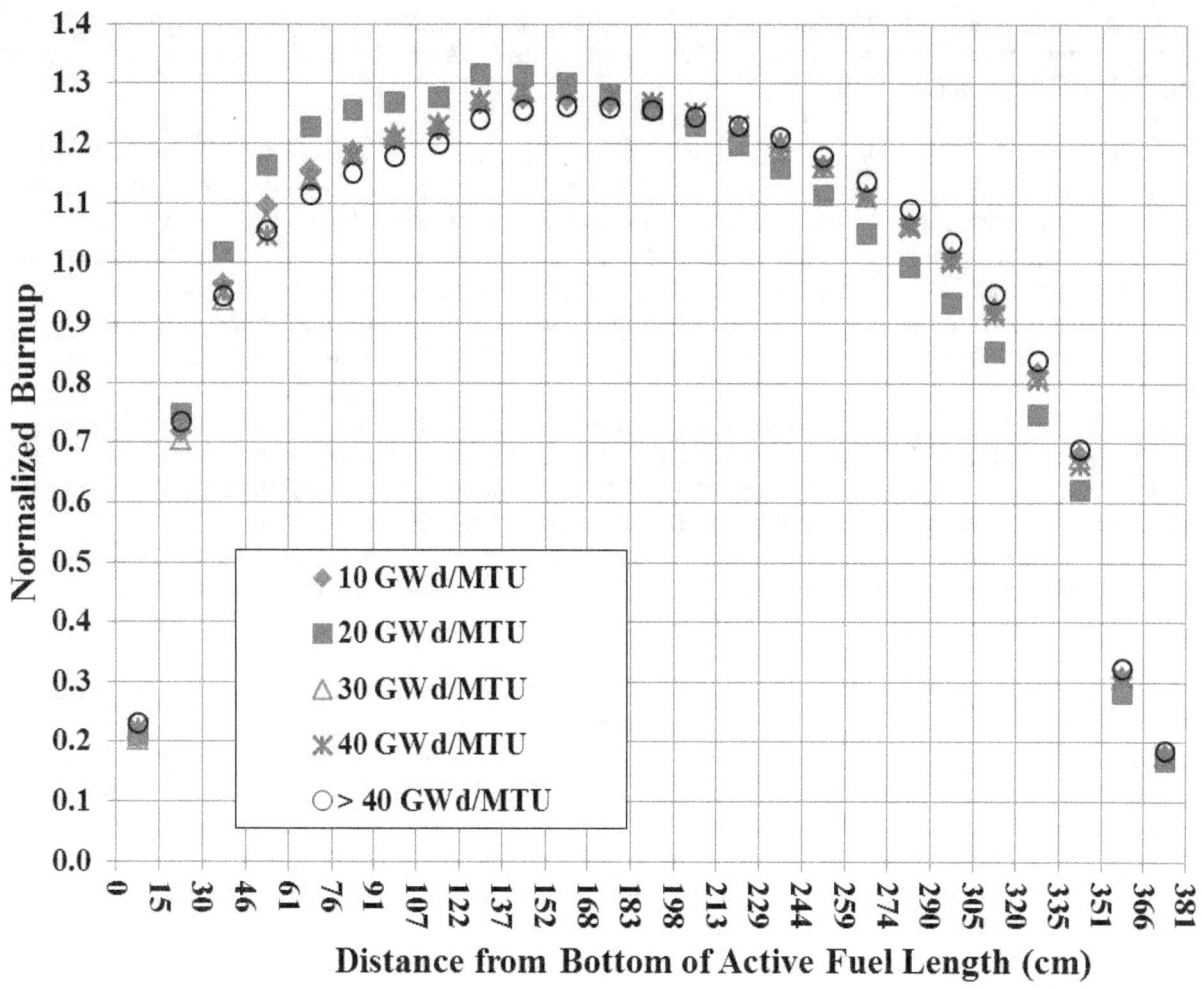

Figure 5. Normalized axial burnup distributions used. These distributions are based on data for assembly B2 from the LaSalle Unit 1 CRC data.

Table 16. Normalized axial burnup distributions used for benchmark problem[†]

Upper bound of axial region, measured from bottom of active fuel (cm)	GBC-68 assembly average burnup (GWd/MTU)				
	10	20	30	40	>40
15.24	0.205	0.210	0.204	0.223	0.230
30.48	0.719	0.747	0.706	0.732	0.735
45.72	0.964	1.018	0.939	0.953	0.945
60.96	1.095	1.164	1.067	1.046	1.054
76.2	1.156	1.229	1.140	1.138	1.114
91.44	1.187	1.256	1.186	1.180	1.151
106.68	1.207	1.269	1.216	1.209	1.178
121.92	1.224	1.278	1.233	1.231	1.200
137.16	1.268	1.316	1.274	1.272	1.240
152.4	1.275	1.313	1.290	1.287	1.257
167.64	1.274	1.301	1.289	1.288	1.262
182.88	1.267	1.282	1.279	1.281	1.261
198.12	1.255	1.258	1.266	1.269	1.255
213.36	1.239	1.230	1.248	1.253	1.246
228.6	1.219	1.198	1.225	1.230	1.231
243.84	1.194	1.160	1.196	1.201	1.210
259.08	1.162	1.115	1.161	1.163	1.180
274.32	1.113	1.051	1.112	1.112	1.137
289.56	1.065	0.995	1.064	1.060	1.091
304.8	1.009	0.935	1.009	1.002	1.036
320.04	0.925	0.854	0.924	0.915	0.949
335.28	0.816	0.748	0.814	0.806	0.838
350.52	0.679	0.622	0.674	0.664	0.691
365.76	0.308	0.282	0.306	0.307	0.324
381.0	0.179	0.168	0.177	0.177	0.186
Assembly B2 average burnup (GWd/MTU)	6.9	19.0	28.9	39.4	44.3

[†] Calculated using axial burnup information for LaSalle Unit 1 assembly B2 [14].

Table 17. k_{eff} and Δk values using axial burnup profiles with and without axial blankets as a function of burnup for fuel with 2 wt% ^{235}U initial enrichment and a 5-year cooling time

Axial model	1 axial zone, no blankets	25 axial zones, no blankets		25 axial zones, with blankets	
Burnup (GWd/MTU)	k_{eff} value*	k_{eff} value*	Δk from 1 axial zone	k_{eff} value*	Δk from 1 axial zone
Major actinides (nuclide set 1, see Table 6)					
0	0.77303	0.77303	0.00000	0.77280	-0.00023
10	0.73359	0.73443	0.00083	0.73020	-0.00339
20	0.68232	0.70575	0.02343	0.69352	0.01120
30	0.64069	0.67723	0.03654	0.65515	0.01446
40	0.61353	0.65838	0.04485	0.63174	0.01821
50	0.59861	0.63940	0.04080	0.61071	0.01210
60	0.59202	0.62591	0.03389	0.59854	0.00652
Actinides and major fission products (nuclide set 2, see Table 6)					
0	0.77303	0.77303	0.00000	0.77280	-0.00023
10	0.70289	0.71071	0.00781	0.70272	-0.00018
20	0.63560	0.67909	0.04349	0.65976	0.02416
30	0.58227	0.64667	0.06440	0.61588	0.03362
40	0.54625	0.62452	0.07827	0.58839	0.04215
50	0.52400	0.60162	0.07762	0.56190	0.03790
60	0.51158	0.58454	0.07296	0.54452	0.03294

*0.0001 Δk Monte Carlo uncertainty for all calculations

Table 18. k_{eff} and Δk values using axial burnup profiles with and without axial blankets as a function of burnup for fuel with 3 wt% ^{235}U initial enrichment and a 5-year cooling time

Axial model	1 axial zone, no blankets	25 axial zones, no blankets		25 axial zones, with blankets	
Burnup (GWd/MTU)	k_{eff} value*	k_{eff} value*	Δk from 1 axial zone	k_{eff} value*	Δk from 1 axial zone
Major actinides (nuclide set 1, see Table 6)					
0	0.87624	0.87624	0.00000	0.87609	-0.00015
10	0.83328	0.83313	-0.00014	0.82807	-0.00520
20	0.77877	0.80239	0.02362	0.78843	0.00966
30	0.72498	0.77046	0.04548	0.74321	0.01822
40	0.67821	0.74872	0.07051	0.71175	0.03355
50	0.64214	0.72533	0.08319	0.67970	0.03756
60	0.61908	0.70794	0.08887	0.65706	0.03798
Actinides and major fission products (nuclide set 2, see Table 6)					
0	0.87624	0.87624	0.00000	0.87609	-0.00015
10	0.80168	0.80782	0.00614	0.79860	-0.00309
20	0.73009	0.77331	0.04321	0.75184	0.02175
30	0.66292	0.73822	0.07530	0.70063	0.03771
40	0.60594	0.71349	0.10755	0.66570	0.05977
50	0.56272	0.68704	0.12432	0.62973	0.06701
60	0.53422	0.66726	0.13304	0.60486	0.07064

*0.0001 Δk Monte Carlo uncertainty for all calculations

Table 19. k_{eff} and Δk values using axial burnup profiles with and without axial blankets as a function of burnup for fuel with 4 wt% ^{235}U initial enrichment and a 5-year cooling time

Axial model	1 axial zone, no blankets	25 axial zones, no blankets		25 axial zones, with blankets	
Burnup (GWd/MTU)	k_{eff} value*	k_{eff} value*	Δk from 1 axial zone	k_{eff} value*	Δk from 1 axial zone
Major actinides (nuclide set 1, see Table 6)					
0	0.94283	0.94283	0.00000	0.94243	-0.00040
10	0.90380	0.90193	-0.00187	0.89793	-0.00587
20	0.85491	0.87243	0.01753	0.86006	0.00515
30	0.80387	0.84107	0.03720	0.81632	0.01246
40	0.75282	0.81902	0.06620	0.78376	0.03095
50	0.70533	0.79487	0.08955	0.74811	0.04278
60	0.66636	0.77667	0.11030	0.72125	0.05489
Actinides and major fission products (nuclide set 2, see Table 6)					
0	0.94283	0.94283	0.00000	0.94243	-0.00040
10	0.87177	0.87504	0.00327	0.86738	-0.00439
20	0.80611	0.84111	0.03500	0.82189	0.01578
30	0.74066	0.80606	0.06540	0.77034	0.02969
40	0.67746	0.78140	0.10394	0.73419	0.05672
50	0.62134	0.75442	0.13308	0.69484	0.07351
60	0.57592	0.73396	0.15804	0.66627	0.09035

*0.0001 Δk Monte Carlo uncertainty for all calculations

Table 20. k_{eff} and Δk values using axial burnup profiles with and without axial blankets as a function of burnup for fuel with 5 wt% ^{235}U initial enrichment and a 5-year cooling time

Axial model	1 axial zone, no blankets	25 axial zones, no blankets		25 axial zones, with blankets	
Burnup (GWd/MTU)	k_{eff} value*	k_{eff} value*	Δk from 1 axial zone	k_{eff} value*	Δk from 1 axial zone
Major actinides (nuclide set 1, see Table 6)					
0	0.99048	0.99048	0.00000	0.99009	-0.00039
10	0.95594	0.95276	-0.00318	0.95004	-0.00590
20	0.91296	0.92483	0.01187	0.91485	0.00189
30	0.86791	0.89491	0.02699	0.87404	0.00612
40	0.82040	0.87362	0.05323	0.84263	0.02224
50	0.77202	0.85008	0.07805	0.80719	0.03517
60	0.72573	0.83194	0.10621	0.77949	0.05377
Actinides and major fission products (nuclide set 2, see Table 6)					
0	0.99048	0.99048	0.00000	0.99009	-0.00039
10	0.92403	0.92513	0.00111	0.91873	-0.00529
20	0.86466	0.89227	0.02760	0.87518	0.01052
30	0.80458	0.85770	0.05312	0.82621	0.02163
40	0.74412	0.83398	0.08987	0.79048	0.04636
50	0.68544	0.80740	0.12197	0.75103	0.06559
60	0.63115	0.78735	0.15620	0.72081	0.08966

*0.0001 Δk Monte Carlo uncertainty for all calculations

Table 21. Axially dependent design of LaSalle Unit 1 assembly B2

Axial Zone	Zone thickness (cm)	Lattice number[†]	Number of fuel rods	Initial enrichment (wt% ^{235}U)	Number of Gd rods	Gd content (wt% Gd_2O_3)
25 (top)	15.24	106	49	0.71	0	0
24	15.24	107	60	0.71	0	0
23	15.24	108	60	3.23	9	3
22	15.24	108	60	3.23	9	3
21	15.24	108	60	3.23	9	3
20	15.24	109	60	3.37	2 & 9[‡]	4 & 3[‡]
19	15.24	109	60	3.37	3 & 9	5 & 3
18	15.24	109	60	3.37	4 & 9	6 & 3
17	15.24	110	60	3.37	9	3
16	15.24	110	60	3.37	9	3
15	15.24	110	60	3.37	9	3
14	15.24	110	60	3.37	9	3
13	15.24	110	60	3.37	9	3
12	15.24	110	60	3.37	9	3
11	15.24	110	60	3.37	9	3
10	15.24	110	60	3.37	9	3
9	15.24	110	60	3.37	9	3
8	15.24	111	60	3.23	5 & 4	4 & 3
7	15.24	111	60	3.23	6 & 4	5 & 3
6	15.24	111	60	3.23	7 & 4	6 & 3
5	15.24	111	60	3.23	8 & 4	7 & 3
4	15.24	111	60	3.23	9 & 4	8 & 3
3	15.24	111	60	3.23	10 & 4	9 & 3
2	15.24	111	60	3.23	11 & 4	10 & 3
1	15.24	107	60	0.71	0	0

[†] Lattice number from Table 3-4 of LaSalle Unit 1 CRC data [14]

[‡] "2 & 9" rods with "4 & 3" gadolinium content means two fuel rods with 4 wt% Gd_2O_3 and nine fuel rods with 3 wt% Gd_2O_3.

4. CONCLUSIONS

This report proposes and documents a computational benchmark model for use as a reference case in the estimation of the additional reactivity margin available from fission products and minor actinides in a BWR burnup-credit storage/transport environment, based on a generic 68 BWR-assembly cask. The proposed benchmark problem was developed to be similar to existing designs for burnup-credit casks, including similar materials and dimensions. While preserving all of the important features, the proposed benchmark problem approximates (or eliminates) nonessential details and proprietary information. The documentation of this computational benchmark includes all of the necessary geometric and material specifications to permit independent evaluations and sufficiently detailed reference solutions to enable meaningful comparisons. This benchmark model will also be used as the base case in future sensitivity studies exploring the impact of varying reactor depletion parameters and burnup-credit analysis modeling approximations.

The purpose of this computational benchmark is to provide a reference configuration to help establish a method for the estimation of the additional reactivity margin and document reference estimations of the additional reactivity margin as a function of initial enrichment, burnup, and cooling time. Calculated k_{eff} values for the benchmark problem are provided as a function of burnup and cooling time for initial enrichments of 2, 3, 4, and 5 wt% ^{235}U. Values are provided for the burnup range of 0 to 60 GWd/MTU, in increments of 10 GWd/MTU, and for cooling times of 0, 5, 10, 20, and 40 years. The individual components of the reactivity reduction associated with (a) the major actinides and (b) the additional nuclides as a function of burnup, cooling time, and initial enrichment are also provided. In addition, results for 2, 3, 4, and 5 wt% ^{235}U initial fuel enrichment with a 5-year post-irradiation cooling time are given for two cases to provide an indication of the potential impact associated with modeling axial burnup profiles with and without axial blankets. The reference estimations were all based on the SCALE 6.1 code package and the SCALE 238-group ENDF/B-VII library.

The reference results are plotted and examined in Appendix D, and in some cases, observations and conclusions are offered. For typical discharge enrichment and burnup combinations, the results show that approximately 70% of the reactivity reduction is due to the major actinides, with the remaining 30% being attributed to the additional nuclides (major fission products and minor actinides). For a given burnup, an increase in the initial enrichment is shown to result in a small decrease in the fractional contribution from the major actinides and a simultaneous increase in the contribution from the additional nuclides. During the time frame of interest, the reactivity reduction associated with the major actinides is shown to increase with cooling time. In contrast, the reactivity reduction associated with the fission products and minor actinides is shown to increase initially with cooling time, but then decrease somewhat in the 5- to 40-year time frame. Finally, the minimum additional reactivity margin available from fission products and minor actinides is quantified for the burnup, initial enrichments, and cooling times considered in this report..

Where applicable, estimates of the reactivity margin for this reference configuration may be compared to those of actual burnup-credit-style casks to provide a check of the design-specific estimates. However, when associating practical meaning to these results, it is important that the reader and potential users of this report understand that this is a computational benchmark, and as such, the reference solutions are based on calculations alone. Although reference solutions are not directly or indirectly based on experimental results, it should be noted that the SCALE depletion (TRITON) and criticality (CSAS5) sequences have been validated using laboratory critical experiments, commercial reactor criticals (CRCs), measured chemical assay data, and reactivity worth measurements with individual fission products important to burnup credit [2].

An additional purpose for development of the GBC-68 computational benchmark model is to provide a reference model to be used in future sensitivity studies exploring the impact of varying fuel depletion parameters and the use of burnup-credit analysis techniques, approximations, and simplifications.

The work described in this report does not include evaluation of the impact of realistic axial variations in BWR fuel assembly designs. Such features include axial zoning of fuel rod initial enrichment, axial zoning of gadolinium content in fuel rods, use of part-length fuel rods, and reduced or natural uranium enrichment blankets on the ends of the fuel. It is possible that the reactivity reduction associated with the major actinides and the additional actinides and fission products will be affected by modeling such axial design features.

Additional work is needed to identify bounding axial burnup distributions appropriate for application to BWR fuel assemblies and to evaluate the interdependence of axial burnup distributions and axial design features. From experience with PWR burnup credit, and as was demonstrated with one example in this report, neglecting the axial burnup distribution may result in conservative k_{eff} value estimates at low assembly average burnup values and in non-conservative estimates of k_{eff} values for assembly average burnups greater than 10 to 20 GWd/MTU. The actual crossover point where using a uniform axial profile changes from conservative to non-conservative will vary with assembly axial design features used and with reactor operations (e.g., control rod usage). It is expected that usage of bounding axial burnup profiles would yield lower estimates for the reactivity margin available from burnup credit nuclides.

Further work is also recommended to quantify the impact of the use of realistic axially dependent reactor operation conditions during fuel depletion calculations. Such conditions include long-term partial insertion of control rods, moderator density axial variation, and fuel temperature axial variation. In the case of the LaSalle Unit 1 CRC data [14], the effective moderator density varies from 0.74 to 0.17 g/cm^3 as the moderator moves from the bottom to the top of the fuel. This will lead to a significantly harder neutron spectrum near the top of the fuel and, consequently, higher plutonium generation in this axial region.

5. REFERENCES

1   J. C. Wagner, *Computational Benchmark for Estimation of Reactivity Margin from Fission Products and Minor Actinides in PWR Burnup Credit*, NUREG/CR-6747 (ORNL/TM-2000/306), U. S. Nuclear Regulatory Commission, Oak Ridge National Laboratory, October 2001.

2   C. V. Parks, M. D. DeHart and J. C. Wagner, *Review and Prioritization of Technical Issues Related to Burnup Credit for LWR Fuel*, NUREG/CR-6665 (ORNL/TM-1999/303), U. S. Nuclear Regulatory Commission, Oak Ridge National Laboratory, February 2000.

3   *SCALE: A Comprehensive Modeling and Simulation Suite for Nuclear Safety Analysis and Design*, ORNL/TM-2005/39, Version 6.1, Oak Ridge National Laboratory, Oak Ridge, Tennessee, June 2011. Available from Radiation Safety Information Computational Center at Oak Ridge National Laboratory as CCC-785.

4   *Nuclear Criticality Safety in Operations with Fissionable Material Outside Reactors*, ANSI/ANS-8.1-1998 (R2007).

5   *Validation of Neutron Transport Methods for Nuclear Criticality Safety Calculations*, ANSI/ANS-8.24-2007.

6   *Burnup Credit for LWR Fuel*, ANSI/ANS-8.27-2008.

7   *Standard Review Plan for Dry Cask Storage Systems*, NUREG-1536, U.S. NRC, Washington, D.C., January 1997.

8   *Standard Review Plan for Transportation Packages for Spent Nuclear Fuel – Draft Report for Comment*, NUREG-1617, U.S. NRC, Washington, D.C., March 1998.

9   J. C. Wagner, M. DeHart, and B. L. Broadhead, *Investigation of Burnup Credit Modeling Issues Associated with BWR Fuel*, ORNL/TM-1999/193, Oak Ridge National Laboratory, October 2000.

10 K. W. Cummings and S. E. Turner, "Design of Wet Storage Racks for Spent BWR Fuel," *Proc. 2001 ANS Embedded Topical Meeting on Practical Implementation of Nuclear Criticality Safety*, American Nuclear Society, Reno, NV, 2001.

11 C. Casado, J. Sabater, and J. Serrano, "Peak Reactivity Characterization and Isotopic Inventory Calculations for BWR Criticality Applications," *International Workshop on Advances in Applications of Burnup Credit for Spent Fuel Storage, Trasnport, Reprocessing, and Disposition*, International Atomic Energy Agency, Cordoba, Spain, 2009.

12 Spent Fuel Project Office Interim Staff Guidance – 8, Revision 0, "Limited Burnup Credit," U.S. Nuclear Regulatory Commission, May 1999.

13   R. Bahney and T. Doering, Interoffice Correspondence to S. Saterlie; R. Memory; and J. Stringer on *Waste Package Sizes Efficiencies and Weights, LV.WP.RHB.05/93-086,* TRW Environmental Safety Systems Inc., Las Vegas, NV, 1993.

14   D. P. Henderson, *Summary Report of Commercial Reactor Criticality Data for LaSalle Unit 1,* B00000000-01717-5705-00138 REV 00, CRWMS/M&O, Las Vegas, NV, September 1999. B0000000001717.

15   J. J. Duderstadt and L. J. Hamilton, *Nuclear Reactor Analysis,* John Wiley & Sons, 1976.

16   *Topical Report on Actinide-Only Burnup Credit for PWR Spent Nuclear Fuel Packages,* DOE/RW-0472, Rev. 2, U.S. Department of Energy, September 1998.

17   G. Radulescu, I. C. Gauld, G. Ilas, and J. C. Wagner, *An Approach for Validating Actinide and Fission Product Burnup Credit Criticality Safety Analyses – Isotopic Composition Predictions,* NUREG/CR-7108 (ORNL/TM-2011/509), U.S. Nuclear Regulatory Commission, Oak Ridge National Laboratory, May 2012.

18   J. M. Scaglione, D. E. Mueller, J. C. Wagner, and W. J. Marshall, *An Approach for Validating Actinide and Fission Product Burnup Credit Criticality Safety Analyses – Criticality (keff)Predictions,* NUREG/CR-7109 (ORNL/TM-2011/514), U.S. Nuclear Regulatory Commission, Oak Ridge National Laboratory, May 2012.

19   J. C. Wagner, M. D. DeHart, and C. V. Parks, *Recommendations for Addressing Axial Burnup in PWR Burnup Credit Analyses,* NUREG/CR-6801 (ORNL/TM-2001/273), U.S. Nuclear Regulatory Commission, Oak Ridge National Laboratory, March 2003.

APPENDIX A. TRITON INPUT FILE EXAMPLE

(Used to generate ORIGEN-ARP Libraries)

```
=t-depl          parm=(addnux=4)
BWR GE 10x10
v7-238
' ------------------------------------------------------------------
'  template to generate libraries for ORIGEN
'  parameters are: 5.0                 - wt% enrichment U-235
'                  95.0                - wt% U-238
'                  0.8                 - coolant density (g/cc)
'                  ge14_uni_rod_5wo_0.8.lib - name of ORIGEN library created
' ------------------------------------------------------------------
' Mixture data
' ------------------------------------------------------------------
read comp
' fuel 1 (no gadolinia)
uo2      1 0.95 840 92235 5.0
                      92238 95.0 end
' clad
zirc2    2 1 567 end
' water coolant
h2o      3 den=0.8 1 512 end
' clad
zirc2    6 1 567 end
' water coolant
h2o      7 den=0.8 1 512 end
' channel tube
zirc2    9 1 512 end
' water in water rod and channel
h2o     10 den=0.776 1 512 end
'
'---------------------Control rod blades----------------------------
'
' Material:    B4C absorber
'              ss304 clad  7.94 g/cc  SCALE default density
  ss304 21 den=7.94   1.0000 512 end
  b4c   22 den=1.76   1.0000 512 end
  h2o   23 den=0.776  1.0000 512 end
end comp
' ------------------------------------------------------------------
' Cell data
' ------------------------------------------------------------------
read celldata
latticecell squarepitch  pitch=1.295    3
                         fueld=0.876    1
                         cladd=1.026    2 end
end celldata
' ------------------------------------------------------------------
' Depletion data
' ------------------------------------------------------------------
read depletion
 1
end depletion
' ------------------------------------------------------------------
```

```
' Burn data
' --------------------------------------------------------------
read burndata
 power=40  burn=1e-15 down=0  end
 power=40  burn=4 down=0  end
 power=40  burn=71 down=0  end
 power=40  burn=75 down=0  end
 power=40  burn=75 down=0  end
 power=40  burn=75 down=0  end
 power=40  burn=75 down=0  end
 power=40  burn=75 down=0  end
 power=40  burn=75 down=0  end
 power=40  burn=75 down=0  end
 power=40  burn=75 down=0  end
 power=40  burn=75 down=0  end
 power=40  burn=75 down=0  end
 power=40  burn=75 down=0  end
 power=40  burn=75 down=0  end
 power=40  burn=75 down=0  end
 power=40  burn=75 down=0  end
 power=40  burn=75 down=0  end
 power=40  burn=75 down=0  end
 power=40  burn=75 down=0  end
 power=40  burn=75 down=0  end
 power=40  burn=75 down=0  end
 power=40  burn=75 down=0  end
 power=40  burn=75 down=0  end
 power=40  burn=75 down=0  end
 power=40  burn=75 down=0  end
 power=40  burn=75 down=0  end
 power=40  burn=75 down=0  end
 power=40  burn=75 down=0  end
 power=40  burn=75 down=0  end
 power=40  burn=75 down=0  end
 power=40  burn=75 down=0  end
end burndata
' --------------------------------------------------------------
' NEWT model data
' --------------------------------------------------------------
read model
BWR GE 10x10
read parm
 fillmix=10 echo=yes drawit=no run=yes
 cmfd=yes xycmfd=4 epsilon=1e-5
end parm
READ materials
  1  1  ! fuel             ! end
  2  1  ! clad             ! end
  3  2  ! coolant          ! end
  9  1  ! channel tube     ! end
 10  2  ! moderator        ! end
 21  1  ! blade ss304      ! end
 22  1  ! blade b4c        ! end
 23  2  ! blade water      ! end
end materials
read geom
unit 1
```

```
com='fuel pin'
 cylinder 10 .438
 cylinder 20 .513
 cuboid   40  4p0.6475
 media 1 1 10
 media 2 1 20 -10
 media 3 1 40 -20
 boundary  40 4 4
unit 3
com='water rod'
 cylinder 10 1.1605
 cylinder 20 1.2605
 media 10 1 10
 media  9 1 20 -10
 boundary 20 8 8
'-------------------------------------------------------------------
' control blade geometry
'-------------------------------------------------------------------
' - GE OEM B4C blade
' - adjusted so that the poison tubes are evenly
'   dispersed in the wing of the blade
unit 90
 com='control poison tube - north wing'
 cylinder 1 0.2108 chord -y=0
 cylinder 2 0.2794 chord -y=0
 cuboid   3 2p0.2876 0 -0.30226
 media 22 1 1
 media 21 1 2 -1
 media 23 1 3 -2
 boundary 3 2 1
'
unit 91
 com='control poison tube - north wing - end tube'
 cylinder 1 0.2108 chord -y=0 chord -x=0
 cylinder 2 0.2794 chord -y=0 chord -x=0
 cuboid   3 0 -0.2876 0 -0.30226
 media 22 1 1
 media 21 1 2 -1
 media 23 1 3 -2
 boundary 3 1 1
'
unit 92
 com='control poison tube - west wing'
 cylinder 1 0.2108 chord +x=0
 cylinder 2 0.2794 chord +x=0
 cuboid   3 0.30226 0 2p0.2876
 media 22 1 1
 media 21 1 2 -1
 media 23 1 3 -2
 boundary 3 1 2
unit 93
 com='control poison tube - west wing - end tube'
 cylinder 1 0.2108 chord +x=0 chord +y=0
 cylinder 2 0.2794 chord +x=0 chord +y=0
 cuboid   3 0.30226 0 0.2876 0
 media 22 1 1
 media 21 1 2 -1
```

```
media 23 1 3 -2
boundary 3 1 1
'
' note: the extra chords on the outer cylinder are to prevent tracer errors -
' the tips of the
'        'circle' (polygon) stick out of the unit boundary
unit 94
 com='control blade wing end tube and rounded edge'
 cylinder 1  0.2108 chord +x=0 chord -y=0  com='b4c'
 cylinder 2  0.2794 chord +x=0 chord -y=0  com='clad'
 cylinder 3 0.30226 chord +x=0 chord -y=0  com='inner wing radius'
 cylinder 4 0.41656 chord +x=0 chord -y=0 chord -x=0.41656 chord +y=-0.41656
com='outer wing radius'
 media 22 1 1
 media 21 1 2 -1
 media 23 1 3 -2
 media 21 1 4 -3
 boundary 4 1 1
'
unit 95
 com='control blade north wing'
 cuboid 1 10.066  0 0 -0.41656
 media 21 1 1
 array 2 1 place 1 1 0.2875 0
 boundary 1
'
unit 96
 com='control blade west wing'
 cuboid 1 0.41656 0 0 -10.066
 media 21 1 1
 array 3 1 place 1 18 0 -0.2875
 boundary 1
unit 97 com='control central support'
 cuboid 1 1.9685 0 0 -0.41656
 media 21 1 1
 boundary 1
unit 98 com='control central support'
 cuboid 1 0.41656 0  0 -1.55194
 media 21 1 1
 boundary 1
'
' end of control blade geometry
'-------------------------------------------------------------------
global unit 5
com='assembly'
 cuboid 51 4p6.475
 cuboid 52 4p6.675
 cuboid 53 4p7.62
 hole 3 origin x=1.295  y=1.295
 hole 3 origin x=-1.295 y=-1.295
 array 1 51 place 1 1 -5.8275 -5.8275
 media 3  1 51
 media 9  1 52 -51
 media 10 1 53 -52
' insert control blade units
' comment out these lines to withdraw the blade
hole 95 origin x=-5.6515  y=7.62  com='north wing'
```

```
hole 96 origin x=-7.62 y=5.6515   com='west wing'
hole 97 origin x=-7.62 y=7.62  com='north support'
hole 98 origin x=-7.62 y=7.20344  com='west support'
hole 94 origin x=4.4145    y=7.62  com='north wing tip'
hole 94 origin x=-7.62 y=-4.4145  com='west wing tip'
'
 boundary 53 36 36
end geom
read array
 ara=1 nux=10 nuy=10
 fill
1 1 1 1 1 1 1 1 1 1
1 1 1 1 1 1 1 1 1 1
1 1 1 1 1 1 1 1 1 1
1 1 1 0 0 1 1 1 1 1
1 1 1 0 0 1 1 1 1 1
1 1 1 1 1 0 0 1 1 1
1 1 1 1 1 0 0 1 1 1
1 1 1 1 1 1 1 1 1 1
1 1 1 1 1 1 1 1 1 1
1 1 1 1 1 1 1 1 1 1
 end fill
' control blade poison tubes array - north wing
 ara=2
 nux=18 nuy=1
 fill
  17r90 91 end fill

' control blade poison tubes array   west wing
 ara=3
 nux=1 nuy=18
 fill
  93 17r92 end fill
end array
read bounds
 all=refl
end bounds
end data
end
' ----------------------------------------------------------------
=shell
mv ft33f001.cmbined $OUTDIR/ge14_uni_rod_5wo_0.8.lib
end
```

APPENDIX B. STARBUCS INPUT FILE EXAMPLE

```
=starbucs
GBC-68 - BUC GE10x10, 5.0 w/o, 60 GWd/MTU, 28Act&FP, cooling time of 40 years
v7-238
read comp
'      - UO2 for fuel pellet
uo2     1 den=10.5216  1 293.0
                    92234    .0445
                    92235    5.0
                    92236    .0230
                    92238    94.9325 end
'      - Zr cladding
zr      2  0  0.04230          293.0   end

'      - water for fuel pin moderator
h2o     3  1                   293.0   end

'      - water for fuel pin pellet/clad gap
h2o     4  1                   293.0   end

'      - water in water tube
h2o     5  1                   293.0   end

'      - zirc for water tube
zr      6  0  0.04230          293.0   end

'      - water outside water tube
h2o     7  1                   293.0   end

'      - water in fuel channel
h2o     8  1                   293.0   end

'      - zr fuel channel
zr      9  0  0.04230          293.0   end

'      - water outside fuel channel
h2o     10 1                   293.0   end

'      - steel cell wall
ss304   11 1                   293.0   end

'      - Boral (9.721e-2 gb10/cm3)
b-10    12 0  5.8465e-03       293.0   end
b-11    12 0  2.3532e-02       293.0   end
c       12 0  7.3447e-03       293.0   end
al      12 0  4.6120e-02       293.0   end

'      - Al plate
al      13 0  0.0602           293.0   end

'      - water in poison panel gap
h2o     14 1                   293.0   end

'      - water outside cells
h2o     15 1                   293.0   end

'      - steel cask body
ss304   16 1                   293.0   end
end comp
```

```
read celldata
  latticecell squarepitch
     fuelr=0.438  1
     gapr= 0.447  4
     cladr=0.513  2
     pitch=1.295  3  end
end celldata
'
read control
  arp=rodded_ge14
  axp=1       end
  mod=0.6  end
  nax=1
  nuc=u-234     1.0
      u-235     1.0
      u-236     1.0
      u-238     1.0
      pu-238    1.0
      pu-239    1.0
      pu-240    1.0
      pu-241    1.0
      pu-242    1.0
      am-241    1.0
      am-243    1.0
      np-237    1.0
      mo-95     1.0
      tc-99     1.0
      ru-101    1.0
      rh-103    1.0
      ag-109    1.0
      cs-133    1.0
      sm-147    1.0
      sm-149    1.0
      sm-150    1.0
      sm-151    1.0
      sm-152    1.0
      nd-143    1.0
      nd-145    1.0
      eu-151    1.0
      eu-153    1.0
      gd-155    1.0  end
end control
read hist
  power=30  burn=2000.00  down=14610.00  nlib=30    end
end hist
'
read keno
'
read parm
  gen=10000 nsk=100 npg=11000
  sig=10.e-5
  htm=no
end parm

read geom
'UO2 fuel pin
unit    1
  cylinder 101 1   0.438                              381.0  0
  cylinder   4 1   0.447                              381.0  0
  cylinder   2 1   0.513                              381.0  0
  cuboid     3 1   0.6475 -0.6475  0.6475 -0.6475     381.0  0
```

```
'water tube
unit     2
  cylinder   5 1   1.1605                                    381.0  0
  cylinder   6 1   1.2605                                    381.0  0
  cuboid     7 1   1.295  -1.295   1.295 -1.295              381.0  0

'2x2 array of pins
unit 3
  array 1  0 0 0

'3x3 array of pins
unit 4
  array 2  0 0 0

'2x3 array of pins
unit 5
  array 3  0 0 0

'3x2 array of pins
unit 6
  array 4  0 0 0

'fuel assembly
unit 7
  array 5  0 0 0

' Top Half Horizontal Boral Panel
unit 40
  cuboid 13 1   6.475 -6.475   0.0254   0                    381.0  0
  cuboid 12 1   6.475 -6.475   0.12826  0                    381.0  0

' Right-Hand Side Half Vertical Boral Panel
unit 50
  cuboid 13 1   0.0254  0   6.475 -6.475                     381.0  0
  cuboid 12 1   0.12826 0   6.475 -6.475                     381.0  0

' Bottom Half Horizontal Boral Panel
unit 60
  cuboid 13 1   6.475 -6.475   0 -0.0254                     381.0  0
  cuboid 12 1   6.475 -6.475   0 -0.12826                    381.0  0

' Left-Hand Side Half Vertical Boral Panel
unit 70
  cuboid 13 1    0   -0.0254   6.475 -6.475                  381.0  0
  cuboid 12 1    0   -0.12826  6.475 -6.475                  381.0  0

' Assembly Basket Cell
unit 101
'    assembly
  array 5  -6.475 -6.475 0
'    water inside the channel
  cuboid  8 1   6.70306 -6.70306   6.70306 -6.70306  381.0  0
'    Zr channel
  cuboid  9 1   6.95706 -6.95706   6.95706 -6.95706  381.0  0
'    water between channel and basket SS plate
  cuboid 10 1   7.52173 -7.52173   7.52173 -7.52173  381.0  0
'    SS basket plate is cell width - boral panel thickness
  cuboid 11 1   8.27173 -8.27173   8.27173 -8.27173  381.0  0
  cuboid 14 1   8.4     -8.4       8.4     -8.4      381.0  0
     hole     40      0       8.27173      0
     hole     50   8.27173       0         0
     hole     60      0      -8.27173      0
     hole     70  -8.27173       0         0
```

```
' Top Boral/Basket Plate
unit 110
   cuboid 12 1  6.475    -6.475     0.10287 0          381.0 0
   cuboid 13 1  6.475    -6.475     0.12827 0          381.0 0
   cuboid 14 1  8.27173 -8.27173    0.12827 0          381.0 0
   cuboid 11 1  8.27173 -8.27173    0.87827 0          381.0 0

' Left-Hand Side Boral/Basket Plate
unit 112
   cuboid 12 1  0 -0.10287   6.475    -6.475             381.0 0
   cuboid 13 1  0 -0.12827   6.475    -6.475             381.0 0
   cuboid 14 1  0 -0.12827   7.52172 -7.52172            381.0 0
   cuboid 11 1  0 -0.87827   7.52172 -7.52172            381.0 0

' Right-Hand Side Boral/Basket Plate
unit 113
   cuboid 12 1  0.10287 0   6.475    -6.475              381.0 0
   cuboid 13 1  0.12827 0   6.475    -6.475              381.0 0
   cuboid 14 1  0.12827 0   7.52172 -7.52172             381.0 0
   cuboid 11 1  0.87827 0   7.52172 -7.52172             381.0 0

' 1x3 array
unit 114
   array 7  -8.4 0 0

' 2x1 array
unit 115
   array 8  -16.8 0 0

' Cask Inner Volume
global unit 200
   array 6 -50.4 0 0
   zhemicyl+y 15 1  87.5                          395.76 -15
' holes for side and top assemblies
     hole 115    0       67.2001   0
     hole 114  -58.8001  0         0
     hole 114   58.8001  0         0
     hole 101   75.6002  8.4       0
     hole 101  -75.6002  8.4       0
' Exterior Half Boral Panels
' Top Plates
     hole 110   -8.4     84.0002   0
     hole 110    8.4     84.0002   0
     hole 110  -25.2     67.2002   0
     hole 110   25.2     67.2002   0
     hole 110  -42       67.2002   0
     hole 110   42       67.2002   0
     hole 110  -58.8001 50.4002    0
     hole 110   58.8001 50.4002    0
     hole 110   75.6002 16.8002    0
     hole 110  -75.6002 16.8002    0
' Left-Hand Side Plates
     hole 112  -16.8001 75.6002    0
     hole 112  -50.4002 58.8003    0
     hole 112  -67.2002 25.2003    0
     hole 112  -67.2002 42.0004    0
     hole 112  -84.0003  8.4       0
'    Right-Hand    Side    Plates
     hole 113   16.8001 75.6002    0
     hole 113   50.4002 58.8003    0
     hole 113   67.2002 25.2003    0
     hole 113   67.2002 42.0004    0
```

```
      hole  113    84.0003  8.4      0
'
'     Steel     Cask/Overpack
  zhemicyl+y 16 1  92.5                              400.76 -25
  zhemicyl+y 16 1  97.5                              405.76 -35
  zhemicyl+y 16 1 107.5                              415.76 -45
'
'     Cube     Surrounding     Cask
  cuboid  0 1  108 -108  108 0                       415.76 -45
end geom
'
'     Assembly     Type:     GE     10x10A01
read array
  ara=1 nux=2 nuy=2 nuz=1  fill  f1     end fill
  ara=2 nux=3 nuy=3 nuz=1  fill  f1     end fill
  ara=3 nux=2 nuy=3 nuz=1  fill  f1     end fill
  ara=4 nux=3 nuy=2 nuz=1  fill  f1     end fill
  ara=5 nux=4 nuy=4 nuz=1
      fill   4 5 5 4
             6 3 2 6
             6 2 3 6
             4 5 5 4    end fill
  ara=6 nux=6 nuy=4 nuz=1  fill  f101  end fill
  ara=7 nux=1 nuy=3 nuz=1  fill  f101  end fill
  ara=8 nux=2 nuy=1 nuz=1  fill  f101  end fill
end array

read bounds
  +xb=specular
  -xb=specular
  +yb=specular
  +zb=void
  -zb=void
' the -yb must be kept at specular because only half of cask is modeled
  -yb=specular
end bounds
end data
end keno
end
=shell
  cp sysin2 $INPDIR/$BASENAME.bu.inp
end
```

APPENDIX C. CSAS5 INPUT FILE EXAMPLE

```
=csas5         parm=()
gbc-68 - buc ge10x10, 5.0 w/o, 60 gwd/mtu, 28act&fp, cooling time of 40 years
v7-238n
read comp
'Node[01][01]    gbc-68 - buc ge10x10, 5.0 w/o, 60 gwd/mtu, 28act&fp, cooling
tim
   u-234        101    0  7.6567E-06   293.0 end
   u-235        101    0  1.6455E-04   293.0 end
   u-236        101    0  1.7046E-04   293.0 end
   u-238        101    0  2.1343E-02   293.0 end
  np-237        101    0  2.1855E-05   293.0 end
  pu-238        101    0  7.6774E-06   293.0 end
  pu-239        101    0  1.4696E-04   293.0 end
  pu-240        101    0  7.5564E-05   293.0 end
  pu-241        101    0  5.7785E-06   293.0 end
  pu-242        101    0  2.2062E-05   293.0 end
  am-241        101    0  3.4818E-05   293.0 end
  am-243        101    0  6.2444E-06   293.0 end
  mo-95         101    0  7.7774E-05   293.0 end
  tc-99         101    0  7.7847E-05   293.0 end
  ru-101        101    0  7.6055E-05   293.0 end
  rh-103        101    0  3.9853E-05   293.0 end
  ag-109        101    0  6.7615E-06   293.0 end
  cs-133        101    0  7.9871E-05   293.0 end
  nd-143        101    0  4.8263E-05   293.0 end
  nd-145        101    0  4.3580E-05   293.0 end
  sm-147        101    0  1.4892E-05   293.0 end
  sm-149        101    0  1.2901E-07   293.0 end
  sm-150        101    0  1.8791E-05   293.0 end
  sm-151        101    0  4.0258E-07   293.0 end
  eu-151        101    0  1.4592E-07   293.0 end
  sm-152        101    0  5.7591E-06   293.0 end
  eu-153        101    0  7.4564E-06   293.0 end
  gd-155        101    0  5.7754E-07   293.0 end
   o-16         101    0  4.6949E-02   293.0 end
  zr-90           2    0  2.1763E-02   293.0 end
  zr-91           2    0  4.7461E-03   293.0 end
  zr-92           2    0  7.2545E-03   293.0 end
  zr-94           2    0  7.3517E-03   293.0 end
  zr-96           2    0  1.1844E-03   293.0 end
   o-16           3    0  3.3377E-02   293.0 end
   h-1            3    0  6.6753E-02   293.0 end
   o-16           4    0  3.3377E-02   293.0 end
   h-1            4    0  6.6753E-02   293.0 end
   o-16           5    0  3.3377E-02   293.0 end
   h-1            5    0  6.6753E-02   293.0 end
  zr-90           6    0  2.1763E-02   293.0 end
  zr-91           6    0  4.7461E-03   293.0 end
  zr-92           6    0  7.2545E-03   293.0 end
  zr-94           6    0  7.3517E-03   293.0 end
  zr-96           6    0  1.1844E-03   293.0 end
   o-16           7    0  3.3377E-02   293.0 end
```

```
h-1                 7    0    6.6753E-02    293.0 end
o-16                8    0    3.3377E-02    293.0 end
h-1                 8    0    6.6753E-02    293.0 end
zr-90               9    0    2.1763E-02    293.0 end
zr-91               9    0    4.7461E-03    293.0 end
zr-92               9    0    7.2545E-03    293.0 end
zr-94               9    0    7.3517E-03    293.0 end
zr-96               9    0    1.1844E-03    293.0 end
o-16               10    0    3.3377E-02    293.0 end
h-1                10    0    6.6753E-02    293.0 end
c                  11    0    3.1849E-04    293.0 end
si-28              11    0    1.5701E-03    293.0 end
si-29              11    0    7.9763E-05    293.0 end
si-30              11    0    5.2642E-05    293.0 end
p-31               11    0    6.9469E-05    293.0 end
cr-50              11    0    7.5918E-04    293.0 end
cr-52              11    0    1.4640E-02    293.0 end
cr-53              11    0    1.6599E-03    293.0 end
cr-54              11    0    4.1322E-04    293.0 end
mn-55              11    0    1.7407E-03    293.0 end
fe-54              11    0    3.4542E-03    293.0 end
fe-56              11    0    5.3698E-02    293.0 end
fe-57              11    0    1.2295E-03    293.0 end
fe-58              11    0    1.6393E-04    293.0 end
ni-58              11    0    5.2842E-03    293.0 end
ni-60              11    0    2.0202E-03    293.0 end
ni-61              11    0    8.7463E-05    293.0 end
ni-62              11    0    2.7787E-04    293.0 end
ni-64              11    0    7.0435E-05    293.0 end
c                  12    0    7.3447E-03    293.0 end
b-10               12    0    5.8465E-03    293.0 end
b-11               12    0    2.3532E-02    293.0 end
al-27              12    0    4.6120E-02    293.0 end
al-27              13    0    6.0200E-02    293.0 end
o-16               14    0    3.3377E-02    293.0 end
h-1                14    0    6.6753E-02    293.0 end
o-16               15    0    3.3377E-02    293.0 end
h-1                15    0    6.6753E-02    293.0 end
c                  16    0    3.1849E-04    293.0 end
si-28              16    0    1.5701E-03    293.0 end
si-29              16    0    7.9763E-05    293.0 end
si-30              16    0    5.2642E-05    293.0 end
p-31               16    0    6.9469E-05    293.0 end
cr-50              16    0    7.5918E-04    293.0 end
cr-52              16    0    1.4640E-02    293.0 end
cr-53              16    0    1.6599E-03    293.0 end
cr-54              16    0    4.1322E-04    293.0 end
mn-55              16    0    1.7407E-03    293.0 end
fe-54              16    0    3.4542E-03    293.0 end
fe-56              16    0    5.3698E-02    293.0 end
fe-57              16    0    1.2295E-03    293.0 end
fe-58              16    0    1.6393E-04    293.0 end
ni-58              16    0    5.2842E-03    293.0 end
ni-60              16    0    2.0202E-03    293.0 end
ni-61              16    0    8.7463E-05    293.0 end
ni-62              16    0    2.7787E-04    293.0 end
ni-64              16    0    7.0435E-05    293.0 end
```

```
end comp
read celldata
latticecell  squarepitch  pitch=   1.29500   3 fueld=   0.87600 101 cladd=
1.02600   2
 gapd=   0.89400   4 end
end celldata
'
read param
 res=1000000

  gen=10000 nsk=100 npg=11000
  sig=10.e-5
  htm=no
end parm

read geom
'UO2 fuel pin
unit      1
   cylinder 101 1   0.438                              381.0  0
   cylinder   4 1   0.447                              381.0  0
   cylinder   2 1   0.513                              381.0  0
   cuboid     3 1   0.6475 -0.6475  0.6475 -0.6475  381.0  0

'water tube
unit      2
   cylinder   5 1   1.1605                             381.0  0
   cylinder   6 1   1.2605                             381.0  0
   cuboid     7 1   1.295  -1.295   1.295  -1.295    381.0  0

'2x2 array of pins
unit 3
   array 1  0 0 0

'3x3 array of pins
unit 4
   array 2  0 0 0

'2x3 array of pins
unit 5
   array 3  0 0 0

'3x2 array of pins
unit 6
   array 4  0 0 0

'fuel assembly
unit 7
   array 5  0 0 0

' Top Half Horizontal Boral Panel
unit 40
   cuboid 13 1  6.475 -6.475  0.0254  0          381.0 0
   cuboid 12 1  6.475 -6.475  0.12826 0          381.0 0

' Right-Hand Side Half Vertical Boral Panel
unit 50
   cuboid 13 1  0.0254  0  6.475 -6.475          381.0 0
```

```
    cuboid 12 1   0.12826 0   6.475 -6.475              381.0 0

' Bottom Half Horizontal Boral Panel
unit 60
    cuboid 13 1   6.475 -6.475   0 -0.0254              381.0 0
    cuboid 12 1   6.475 -6.475   0 -0.12826             381.0 0

' Left-Hand Side Half Vertical Boral Panel
unit 70
    cuboid 13 1   0   -0.0254   6.475 -6.475            381.0 0
    cuboid 12 1   0   -0.12826 6.475 -6.475             381.0 0

' Assembly Basket Cell
unit 101
'    assembly
    array 5  -6.475 -6.475 0
'    water inside the channel
    cuboid  8 1   6.70306 -6.70306   6.70306 -6.70306  381.0 0
'    Zr channel
    cuboid  9 1   6.95706 -6.95706   6.95706 -6.95706  381.0 0
'    water between channel and basket SS plate
    cuboid 10 1   7.52173 -7.52173   7.52173 -7.52173  381.0 0
'    SS basket plate is cell width - boral panel thickness
    cuboid 11 1   8.27173 -8.27173 8.27173 -8.27173    381.0 0
    cuboid 14 1   8.4     -8.4      8.4     -8.4        381.0 0
       hole      40      0      8.27173       0
       hole      50    8.27173     0          0
       hole      60      0     -8.27173       0
       hole      70   -8.27173     0          0

' Top Boral/Basket Plate
unit 110
    cuboid 12 1   6.475   -6.475     0.10287 0          381.0 0
    cuboid 13 1   6.475   -6.475     0.12827 0          381.0 0
    cuboid 14 1   8.27173 -8.27173   0.12827 0          381.0 0
    cuboid 11 1   8.27173 -8.27173   0.87827 0          381.0 0

' Left-Hand Side Boral/Basket Plate
unit 112
    cuboid 12 1   0 -0.10287   6.475   -6.475           381.0 0
    cuboid 13 1   0 -0.12827   6.475   -6.475           381.0 0
    cuboid 14 1   0 -0.12827   7.52172 -7.52172         381.0 0
    cuboid 11 1   0 -0.87827   7.52172 -7.52172         381.0 0

' Right-Hand Side Boral/Basket Plate
unit 113
    cuboid 12 1   0.10287 0   6.475   -6.475            381.0 0
    cuboid 13 1   0.12827 0   6.475   -6.475            381.0 0
    cuboid 14 1   0.12827 0   7.52172 -7.52172          381.0 0
    cuboid 11 1   0.87827 0   7.52172 -7.52172          381.0 0

' 1x3 array
unit 114
    array 7  -8.4 0 0

' 2x1 array
unit 115
```

```
   array  8  -16.8 0 0

' Cask Inner Volume
global unit 200
   array  6 -50.4 0 0
   zhemicyl+y 15 1  87.5                                    395.76 -15
' holes for side and top assemblies
      hole  115     0         67.2001    0
      hole  114   -58.8001    0          0
      hole  114    58.8001    0          0
      hole  101    75.6002    8.4        0
      hole  101   -75.6002    8.4        0
' Exterior Half Boral Panels
' Top Plates
      hole  110    -8.4       84.0002    0
      hole  110     8.4       84.0002    0
      hole  110   -25.2       67.2002    0
      hole  110    25.2       67.2002    0
      hole  110   -42         67.2002    0
      hole  110    42         67.2002    0
      hole  110   -58.8001    50.4002    0
      hole  110    58.8001    50.4002    0
      hole  110    75.6002    16.8002    0
      hole  110   -75.6002    16.8002    0
' Left-Hand Side Plates
      hole  112   -16.8001    75.6002    0
      hole  112   -50.4002    58.8003    0
      hole  112   -67.2002    25.2003    0
      hole  112   -67.2002    42.0004    0
      hole  112   -84.0003    8.4        0
'      Right-Hand     Side     Plates
      hole  113    16.8001    75.6002    0
      hole  113    50.4002    58.8003    0
      hole  113    67.2002    25.2003    0
      hole  113    67.2002    42.0004    0
      hole  113    84.0003    8.4        0
'
'      Steel     Cask/Overpack
   zhemicyl+y 16 1  92.5                                    400.76 -25
   zhemicyl+y 16 1  97.5                                    405.76 -35
   zhemicyl+y 16 1 107.5                                    415.76 -45
'
'      Cube     Surrounding     Cask
   cuboid  0 1  108 -108  108 0                             415.76 -45
end geom
'
'      Assembly     Type:     GE     10x10A01
read array
   ara=1 nux=2 nuy=2 nuz=1  fill  f1     end fill
   ara=2 nux=3 nuy=3 nuz=1  fill  f1     end fill
   ara=3 nux=2 nuy=3 nuz=1  fill  f1     end fill
   ara=4 nux=3 nuy=2 nuz=1  fill  f1     end fill
   ara=5 nux=4 nuy=4 nuz=1
      fill   4 5 5 4
             6 3 2 6
             6 2 3 6
             4 5 5 4   end fill
```

```
   ara=6 nux=6 nuy=4 nuz=1  fill  f101  end fill
   ara=7 nux=1 nuy=3 nuz=1  fill  f101  end fill
   ara=8 nux=2 nuy=1 nuz=1  fill  f101  end fill
end array

read bounds
  +xb=specular
  -xb=specular
  +yb=specular
  +zb=void
  -zb=void
' the -yb must be kept at specular because only half of cask is modeled
  -yb=specular
end bounds
end data
end
```

APPENDIX D. DISCUSSION OF REFERENCE RESULTS

D.1 DISCUSSION OF REFERENCE RESULTS

Although the intended scope of this report does not include analysis and interpretation of the results, some limited examination, discussion, and conclusions related to the reference results are offered in this appendix.

To assist in the examination of the reference results, plots of a select set of the reference results are presented for a cooling time of 5 years. Figures D.1 through D.4 plot the k_{eff} values in the GBC-68 cask for the two nuclide sets (refer to Tables D.1 and D.2 for specification of nuclide sets) as a function of burnup for initial enrichments of 2, 3, 4, and 5 wt% ^{235}U, respectively. Although, for consistency, results were generated for burnups up to 60 GWd/MTU for each of the enrichments, it should be noted that typical discharge burnups for fuel assemblies with initial enrichments below 3 wt % ^{235}U are below 40 GWd/MTU. Similarly, for initial enrichments above 3 wt % ^{235}U, typical discharge burnups are greater than 30 GWd/MTU. Thus, one should be cognizant of these facts when examining the results for the various burnup and enrichment combinations. Limited data for actual initial enrichment and discharge burnup combinations for SNF discharged from U.S. reactors through 2002 are available in Ref. D.1.

The reactivity reductions due to the major actinides (set 1 in Table D.1), additional nuclides (set 3 in Table D.2), and all of the nuclides considered (set 2 in Table D.1) are plotted in Figures D.5 through D.8 as a function of burnup for each of the initial enrichments considered. Finally, bar charts showing the individual contributions from (a) the major actinides and (b) the additional nuclides to the total reactivity reduction for 5-year cooling as a function of burnup are plotted in Figures D.9 through D.12. The relative reactivity reductions due to (a) the major actinides and (b) the additional nuclides are similar at low burnups, but quickly diverge for higher burnups. For typical enrichment and discharge burnup combinations, the figures show that approximately 70% of the reactivity reduction is due to the major actinides, with the remaining 30% being attributed to the additional nuclides (major fission products and minor actinides).

To compare the reactivity reductions for various initial enrichments, Figure D.13 shows the reactivity reductions (in terms of Δk values) for the major actinides as a function of burnup from Figures D.5 through D.8 for initial enrichments of 2, 3, 4, and 5 wt % ^{235}U. Similarly, Figure D.14 shows the reactivity reductions for the additional nuclides as a function of burnup from Figures D.5 through D.8 for initial enrichments of 2, 3, 4, and 5 wt % ^{235}U. The variation in these contributions to the reactivity reduction versus initial enrichment is shown to be relatively minor. However, Figure D.14 does reveal that, for a given burnup, an increase in the initial enrichment results in a slight increase in the contribution from the additional nuclides. The individual contributions to the total reactivity reduction from (a) the major actinides and (b) the additional nuclides from Figures D.9 through D.12 are compared for various initial enrichments in Figures D.15 and D.16, respectively.

To examine the effect of cooling time, Figures D.17 and D.18 show the reactivity reductions (in terms of Δk values) as a function of burnup, for an initial enrichment of 4 wt % ^{235}U, for (a) the major actinides and (b) the additional nuclides, respectively. During the time frame of interest to storage and transport, the reactivity reduction associated with the major actinides increases with cooling time, primarily due to the decay of the ^{241}Pu fissile nuclide ($t_{1/2}$ = 14.4 years) and the buildup of the neutron absorber ^{241}Am (from decay of ^{241}Pu). This behavior has been well documented [D.2,D.3]. For the fission products and minor actinides considered as the "additional nuclides" (see Table D.2), the associated reactivity reduction increases initially with cooling time due to the buildup of ^{155}Gd (from ^{155}Eu which decays with $t_{1/2}$ = 4.7 years), but then decreases somewhat due to the decay of ^{151}Sm. In general, however, the reactivity reduction due to the additional nuclides does not vary significantly in the 5-to-40 year time frame.

As there has been a great deal of interest in quantifying the minimum additional reactivity margin available from fission products and minor actinides, Figure D.19 plots the range of calculated reactivity margin (in terms of Δk values) for all of the burnup, initial enrichments, and cooling times considered in this report. Specifically, the range for burnups from 10 to 60 GWd/MTU, initial enrichments of 2, 3, 4, and 5 wt % ^{235}U, and cooling times of 0, 5, 10, 20, and 40 years. In all cases, the minimum values correspond to zero cooling time.

Table D.1. Nuclide sets defined for the benchmark problem analysis

set 1: Major actinides (9 total)									
^{234}U	^{235}U	^{238}U	^{238}Pu	^{239}Pu	^{240}Pu	^{241}Pu	^{242}Pu	^{241}Am	

set 2: Actinides and major fission products (28 total)									
^{234}U	^{235}U	^{236}U	^{238}U	^{238}Pu	^{239}Pu	^{240}Pu	^{241}Pu	^{242}Pu	^{241}Am
^{243}Am	^{237}Np	^{95}Mo	^{99}Tc	^{101}Ru	^{103}Rh	^{109}Ag	^{133}Cs	^{147}Sm	^{149}Sm
^{150}Sm	^{151}Sm	^{152}Sm	^{143}Nd	^{145}Nd	^{151}Eu	^{153}Eu	^{155}Gd		

Table D.2. Nuclides in "set 3," on which the additional reactivity margin available from fission products and minor actinides is based

set 3: Minor actinides and major fission products (19 total)									
^{236}U	^{243}Am	^{237}Np	^{95}Mo	^{99}Tc	^{101}Ru	^{103}Rh	^{109}Ag	^{133}Cs	^{147}Sm
^{149}Sm	^{150}Sm	^{151}Sm	^{152}Sm	^{143}Nd	^{145}Nd	^{151}Eu	^{153}Eu	^{155}Gd	

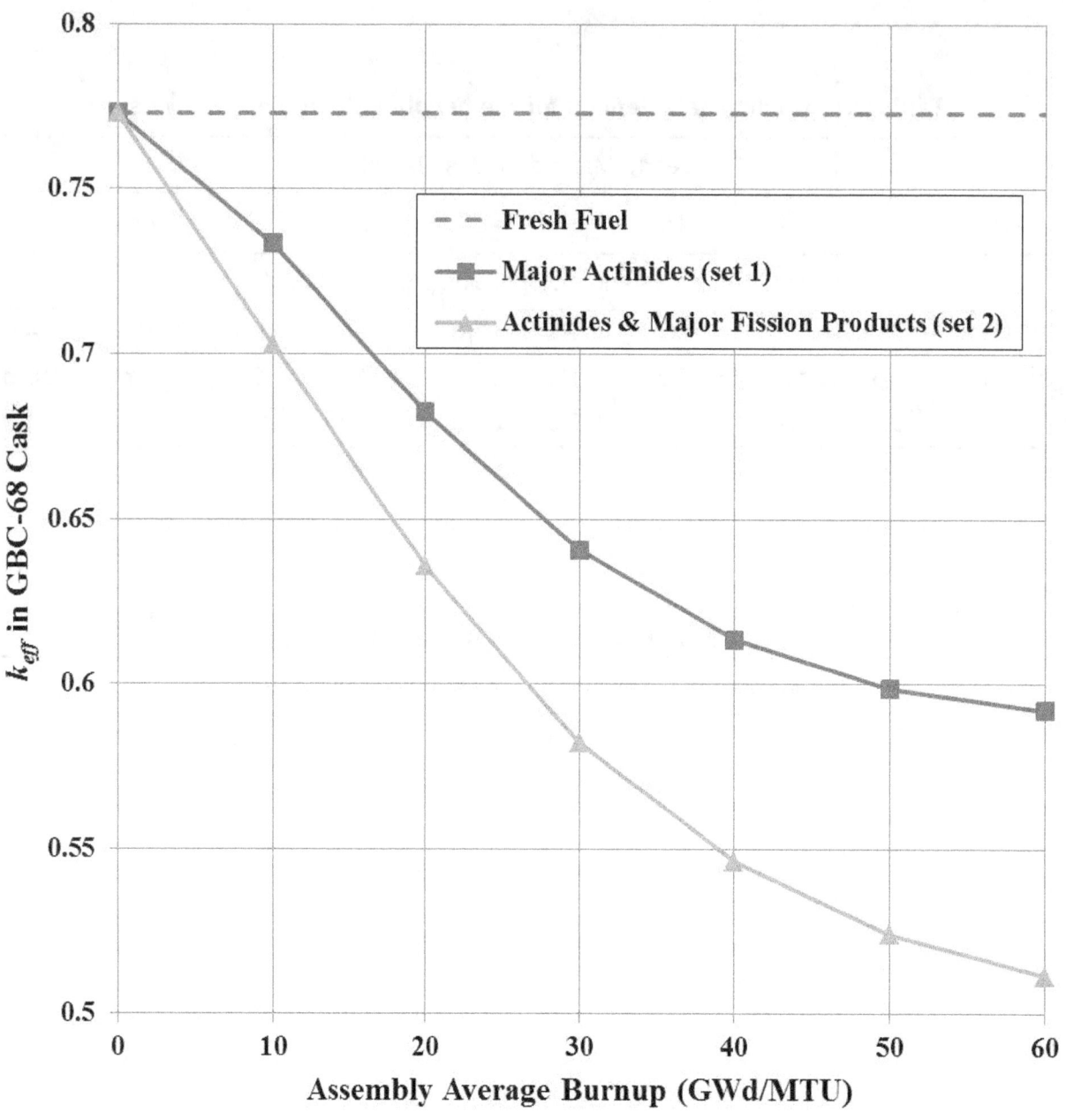

Figure D.1. Values of k_{eff} in the GBC-68 cask as a function of burnup using the different nuclide sets and 5-year cooling time for fuel of 2 wt% ^{235}U initial enrichment.

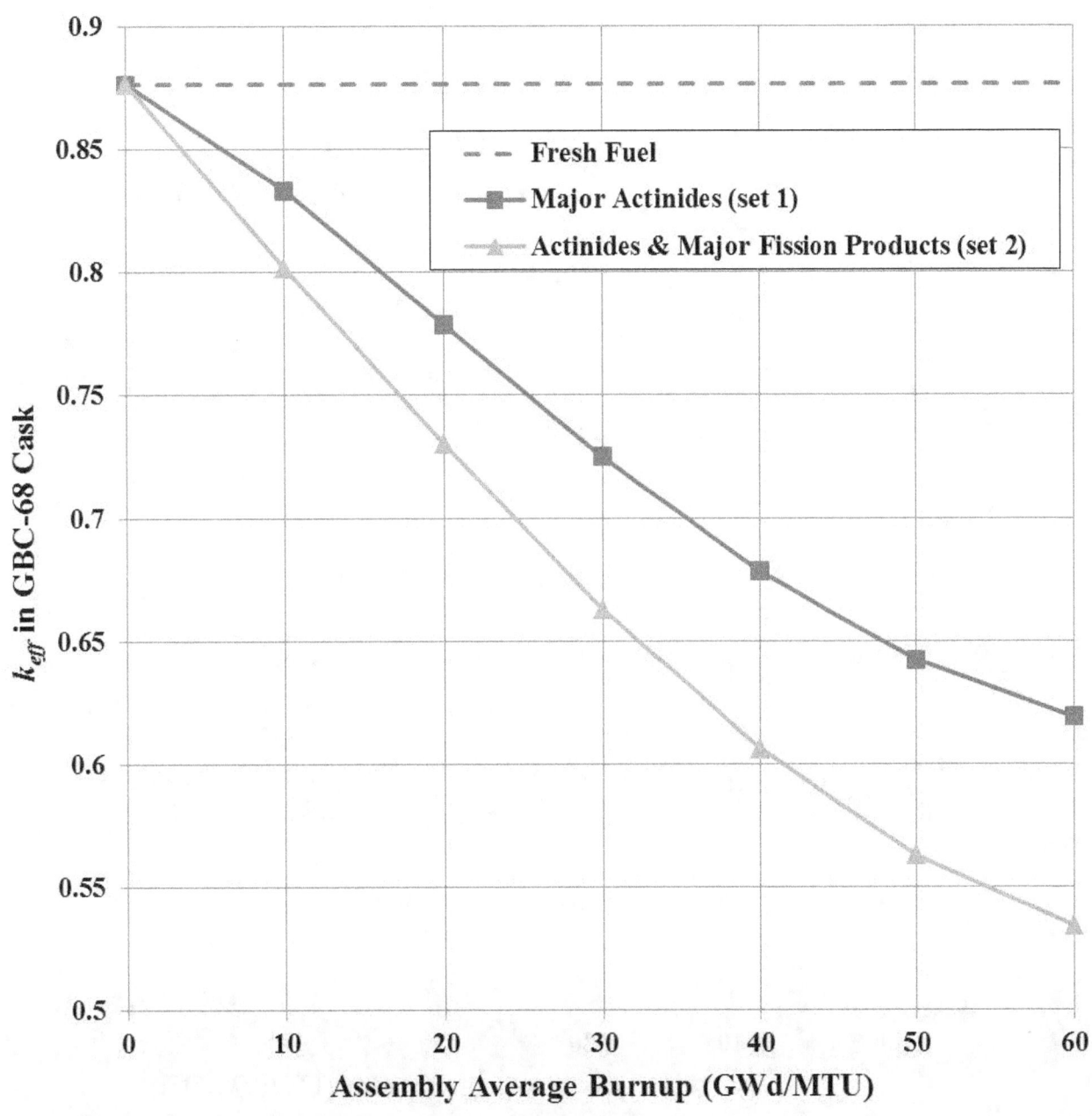

Figure D.2. Values of k_{eff} in the GBC-68 cask as a function of burnup using the different nuclide sets and 5-year cooling time for fuel of 3 wt% ^{235}U initial enrichment.

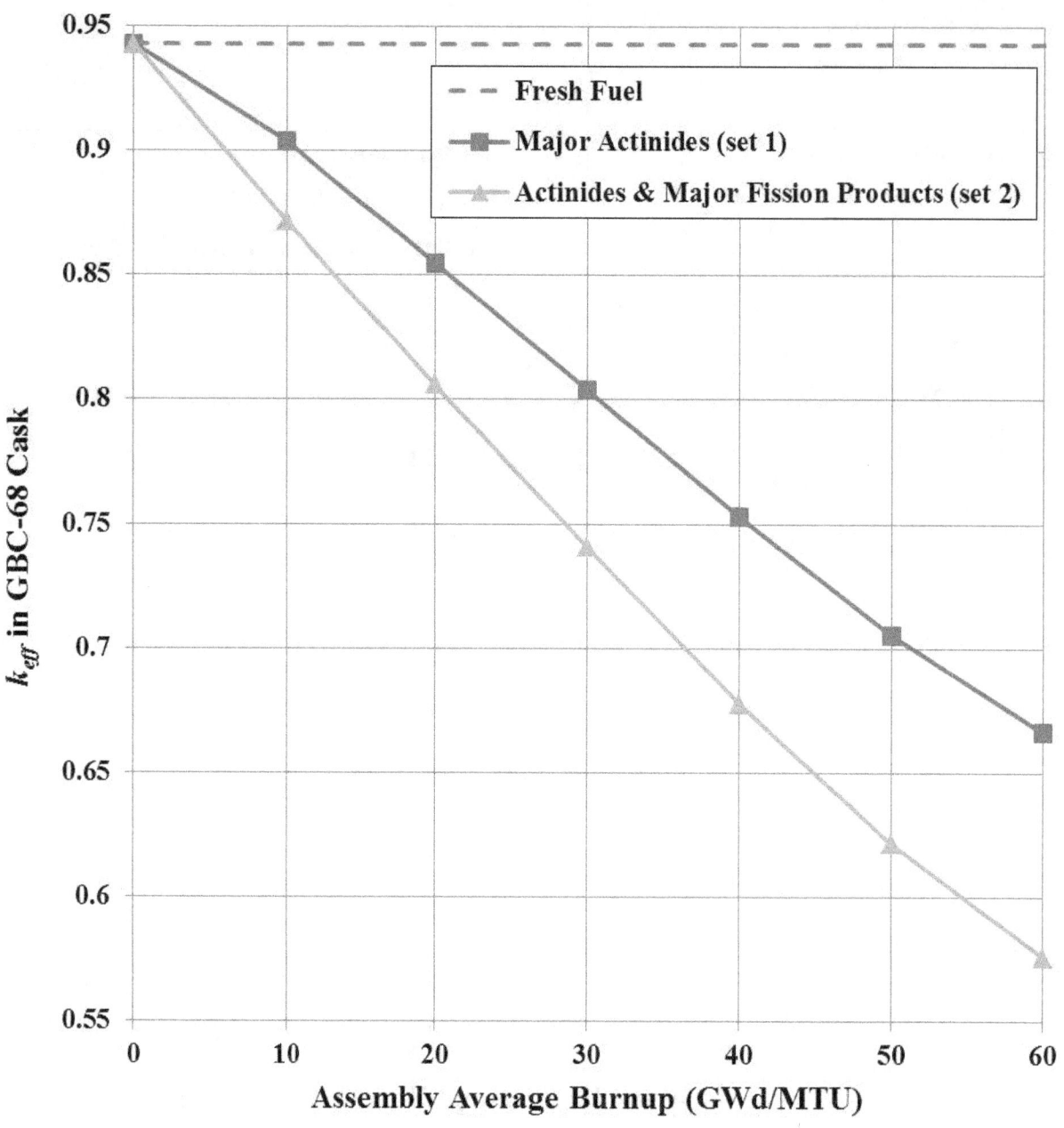

Figure D.3. Values of k_{eff} in the GBC-68 cask as a function of burnup using the different nuclide sets and 5-year cooling time for fuel of 4 wt% ^{235}U initial enrichment.

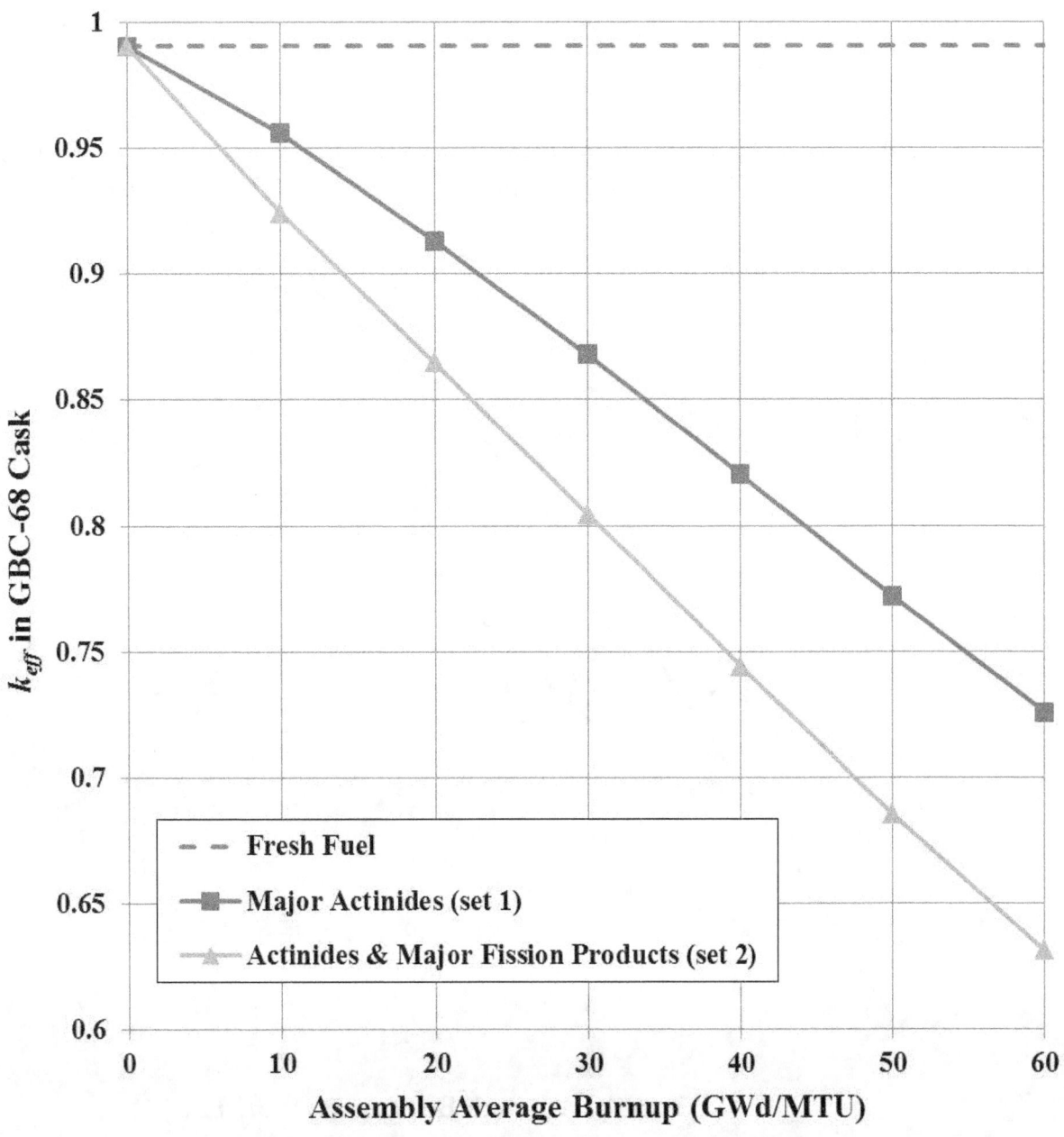

Figure D.4. Values of k_{eff} in the GBC-68 cask as a function of burnup using the different nuclide sets and 5-year cooling time for fuel of 5 wt% ^{235}U initial enrichment.

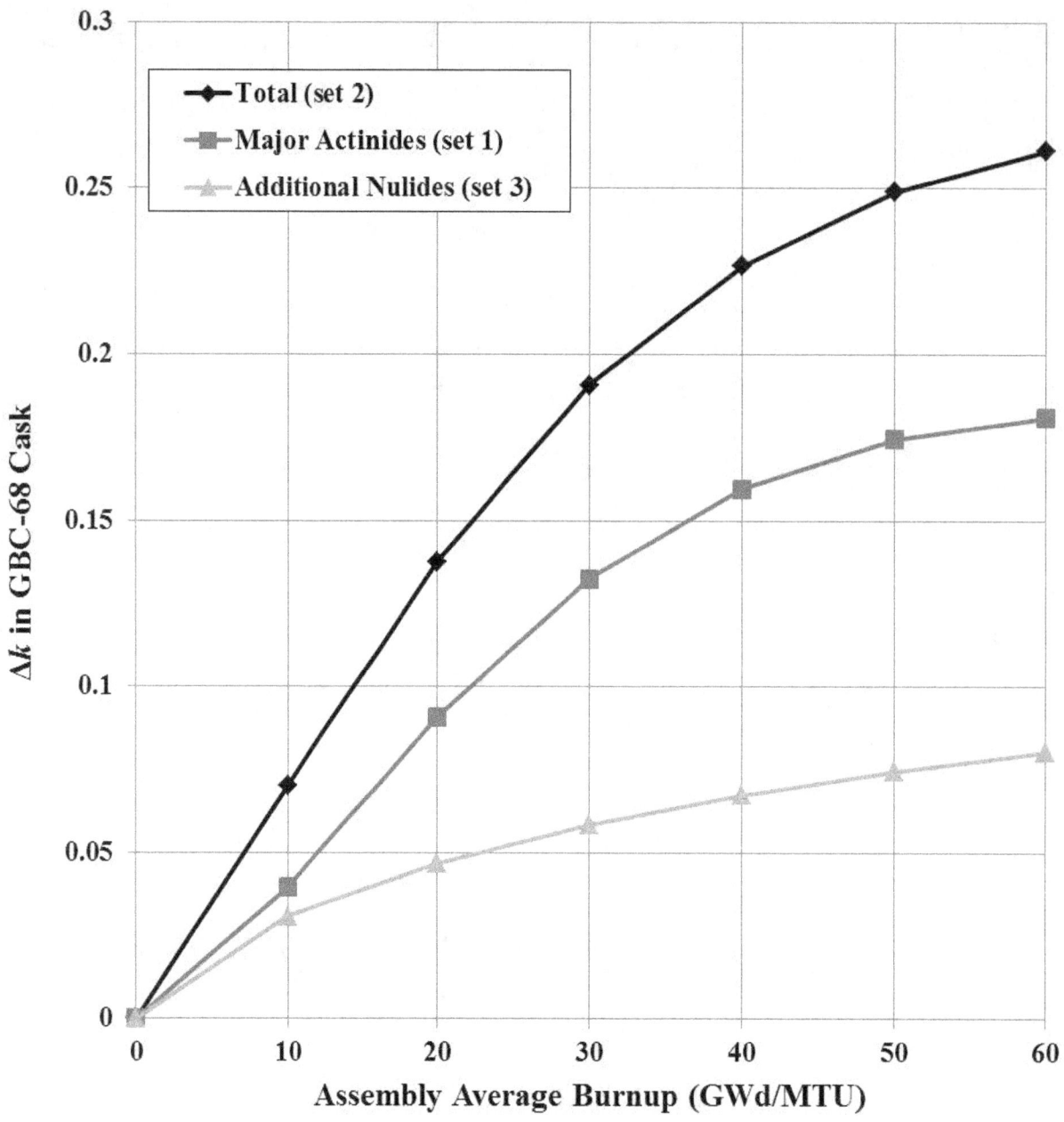

Figure D.5. Δ*k* values (relative to fresh fuel) in the GBC-68 cask as a function of burnup using the different nuclide sets and 5-year cooling time for fuel of 2 wt% ^{235}U initial enrichment.

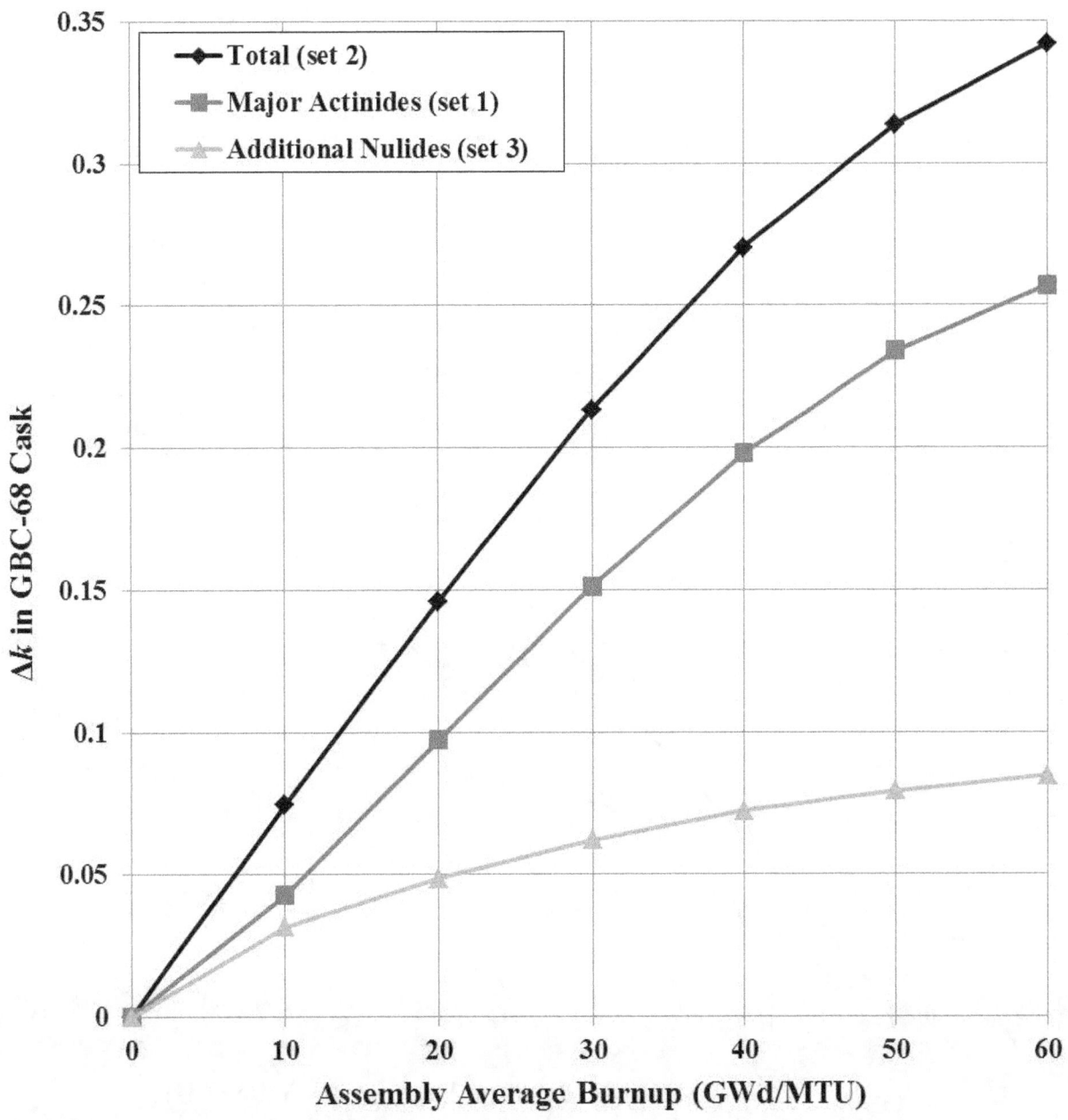

Figure D.6. Δ*k* values (relative to fresh fuel) in the GBC-68 cask as a function of burnup using the different nuclide sets and 5-year cooling time for fuel of 3 wt% ^{235}U initial enrichment.

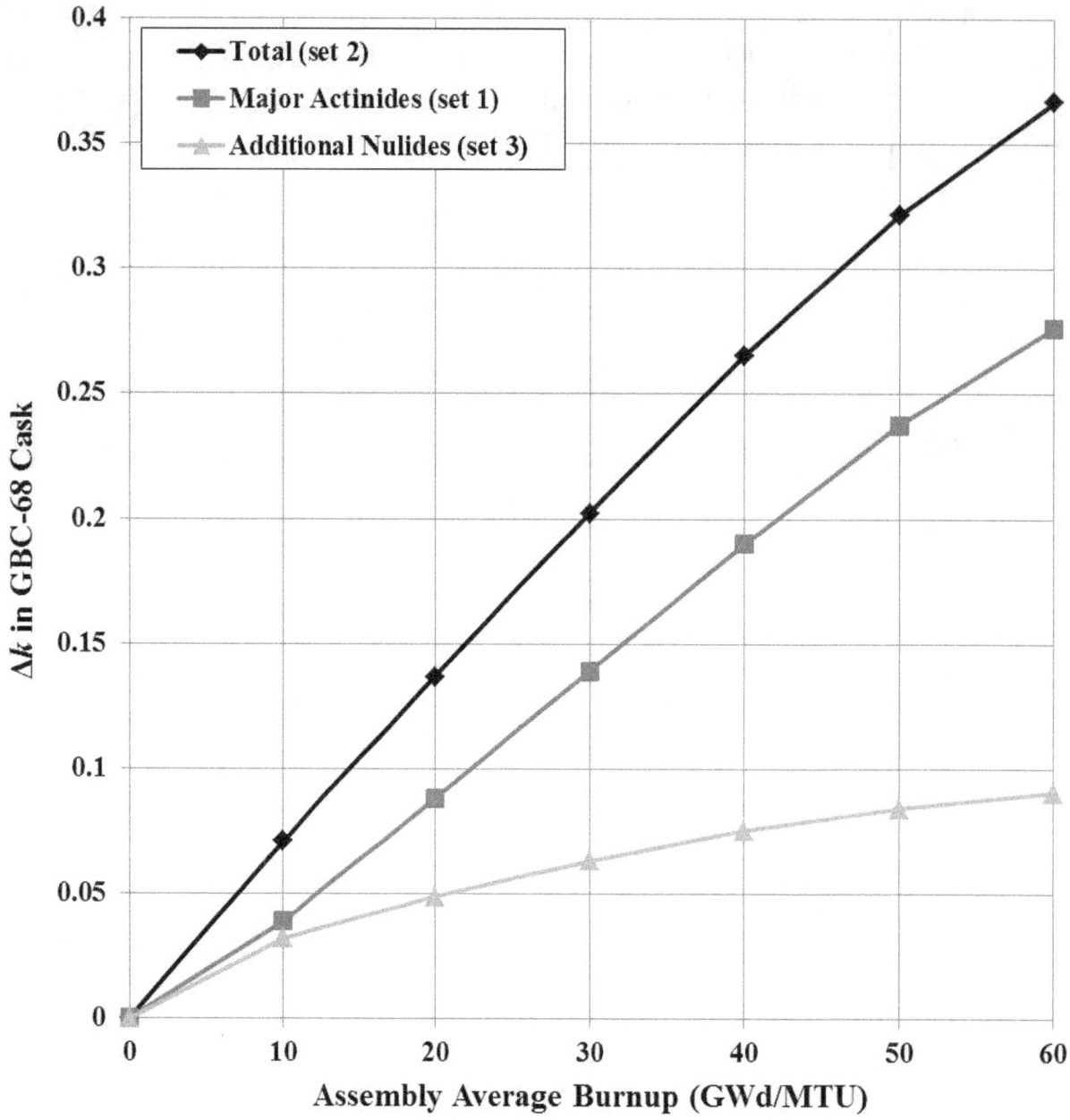

Figure D.7. Δ*k* values (relative to fresh fuel) in the GBC-68 cask as a function of burnup using the different nuclide sets and 5-year cooling time for fuel of 4 wt% ^{235}U initial enrichment.

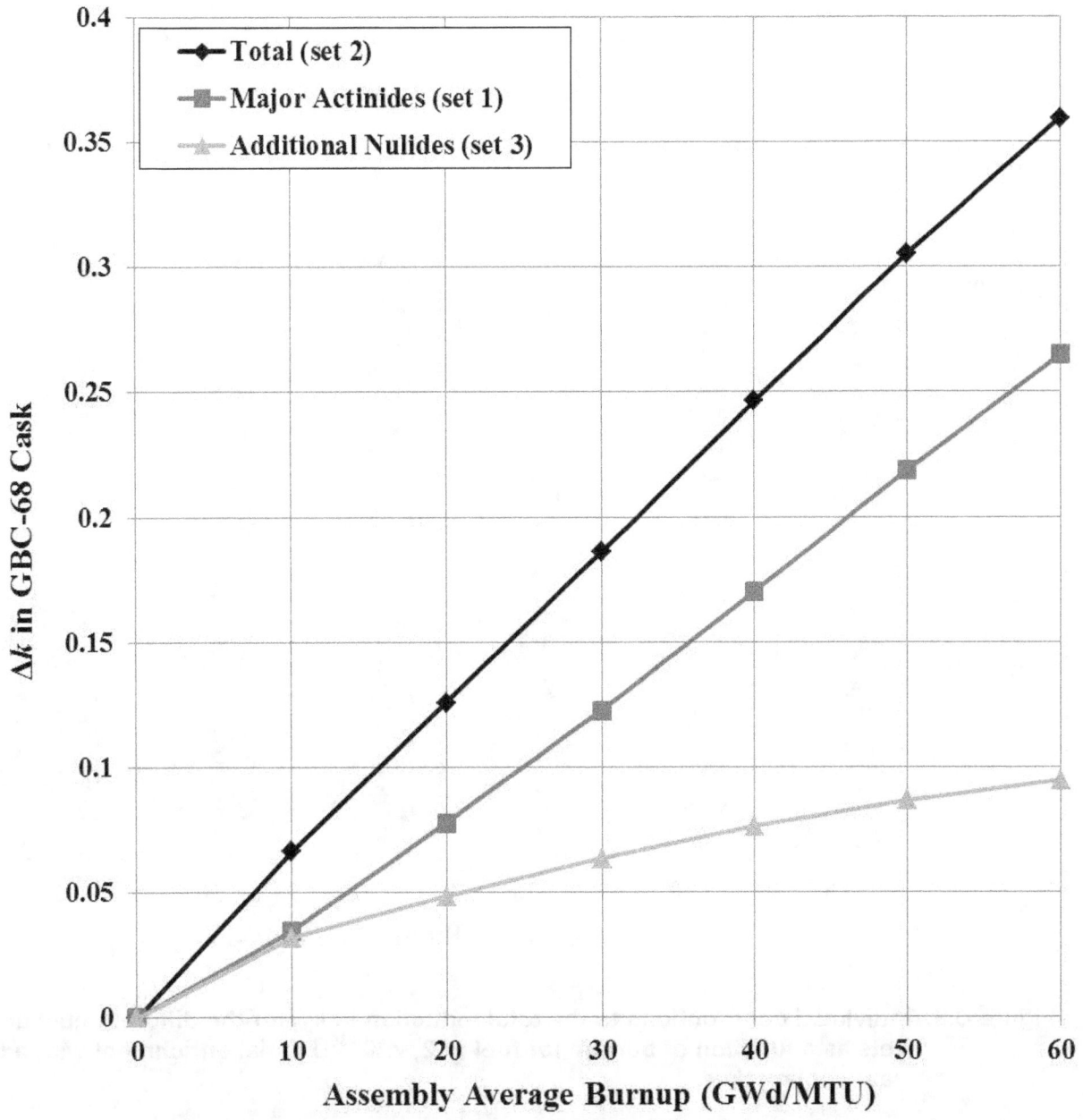

Figure D.8. Δ*k* values (relative to fresh fuel) in the GBC-68 cask as a function of burnup using the different nuclide sets and 5-year cooling time for fuel of 5 wt% ^{235}U initial enrichment.

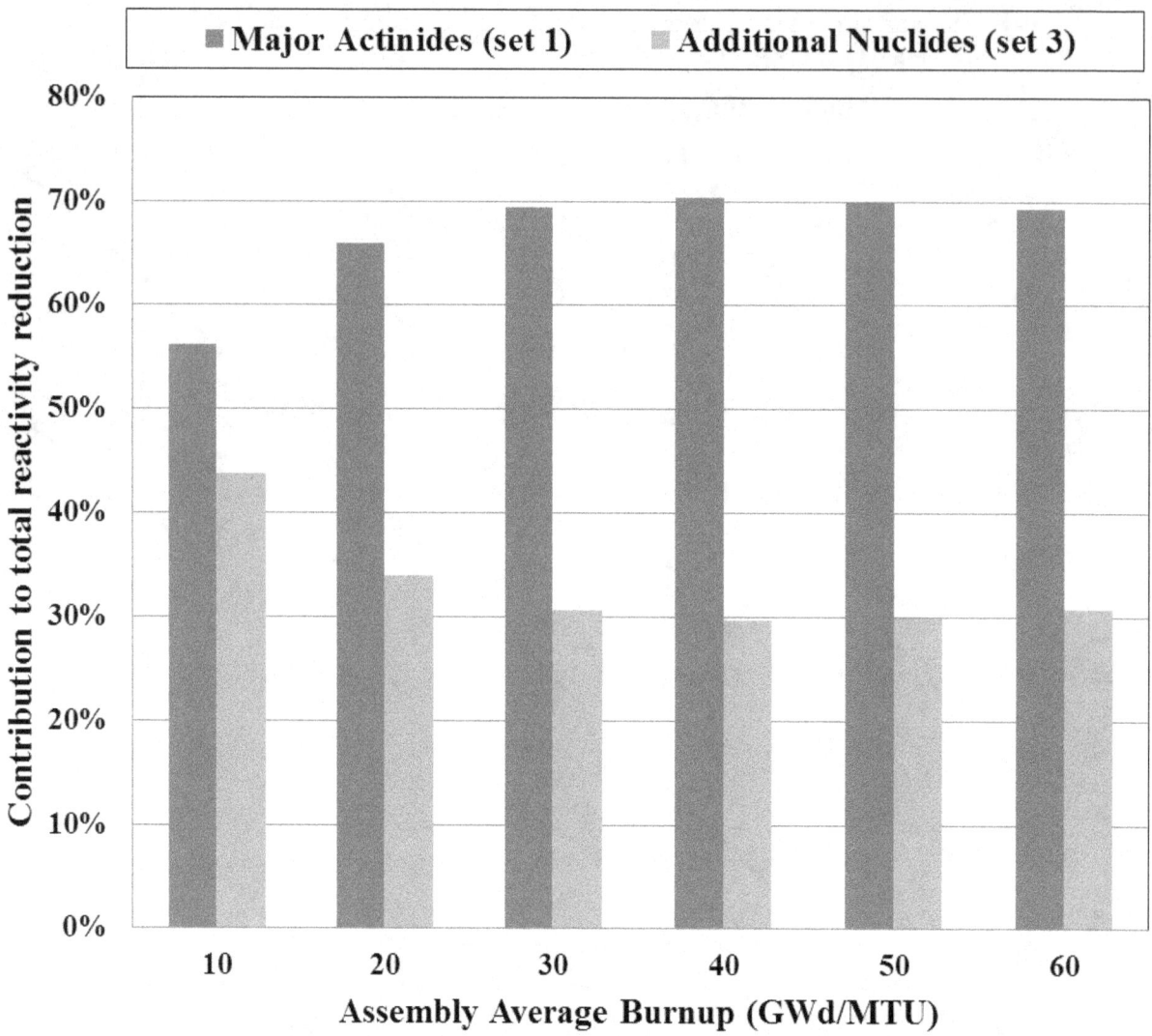

Figure D.9. Individual contributions to the total reduction in k_{eff} for the different nuclide sets as a function of burnup for fuel of 2 wt% ^{235}U initial enrichment with a 5-year cooling time.

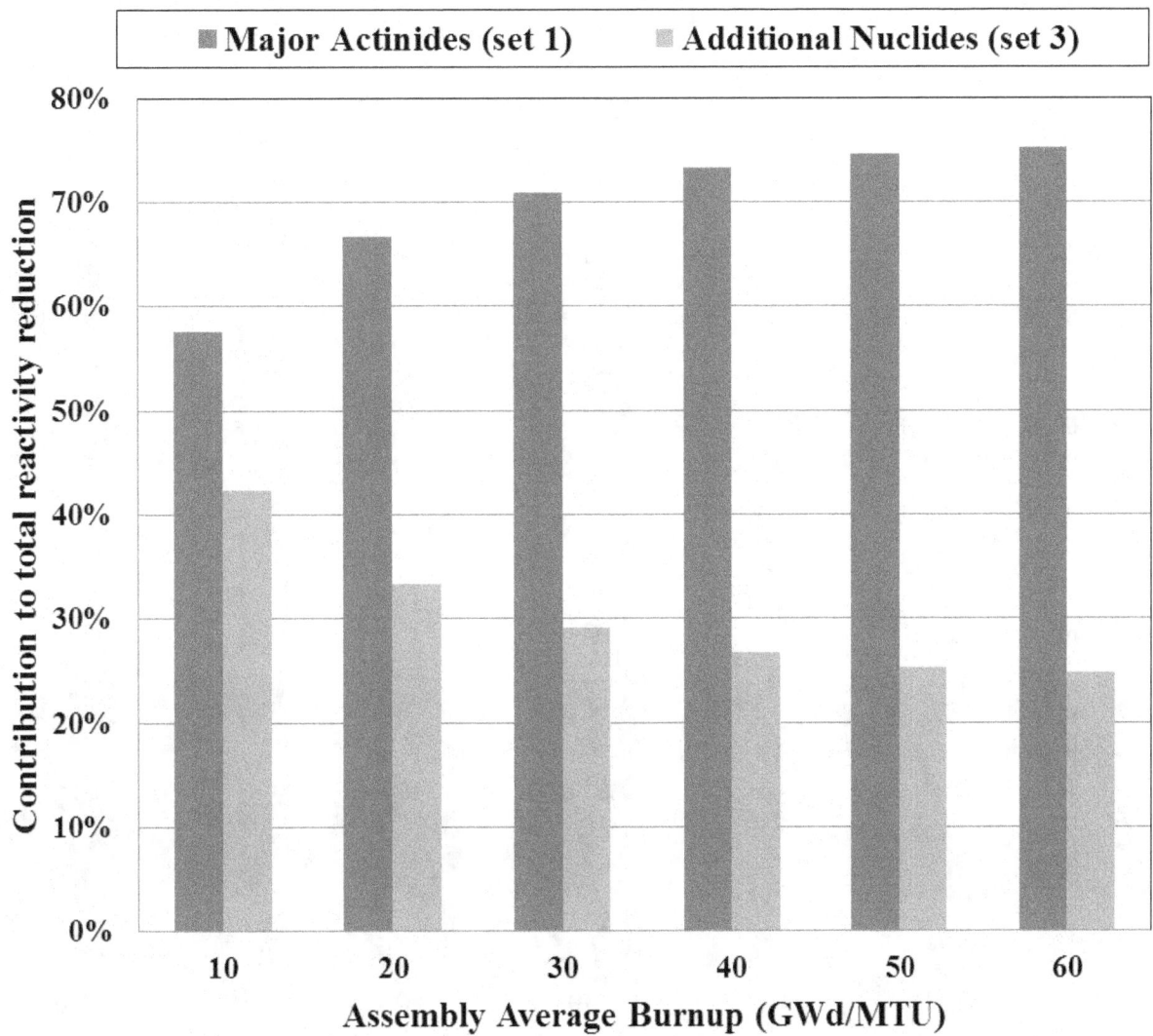

Figure D.10. Individual contributions to the total reduction in k_{eff} for the different nuclide sets as a function of burnup for fuel of 3 wt% ^{235}U initial enrichment with a 5-year cooling time.

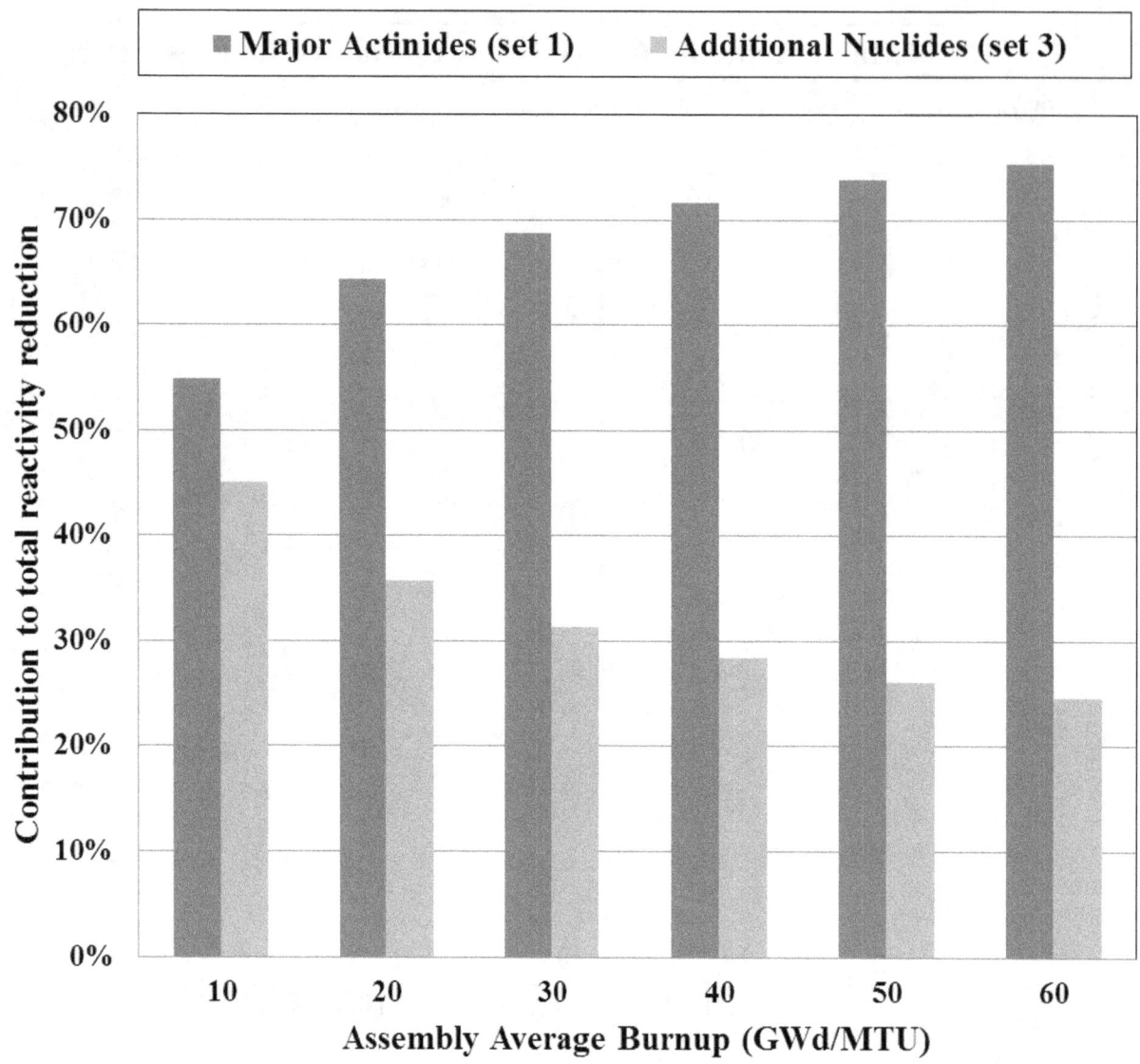

Figure D.11. Individual contributions to the total reduction in k_{eff} for the different nuclide sets as a function of burnup for fuel of 4 wt% ^{235}U initial enrichment with a 5-year cooling time.

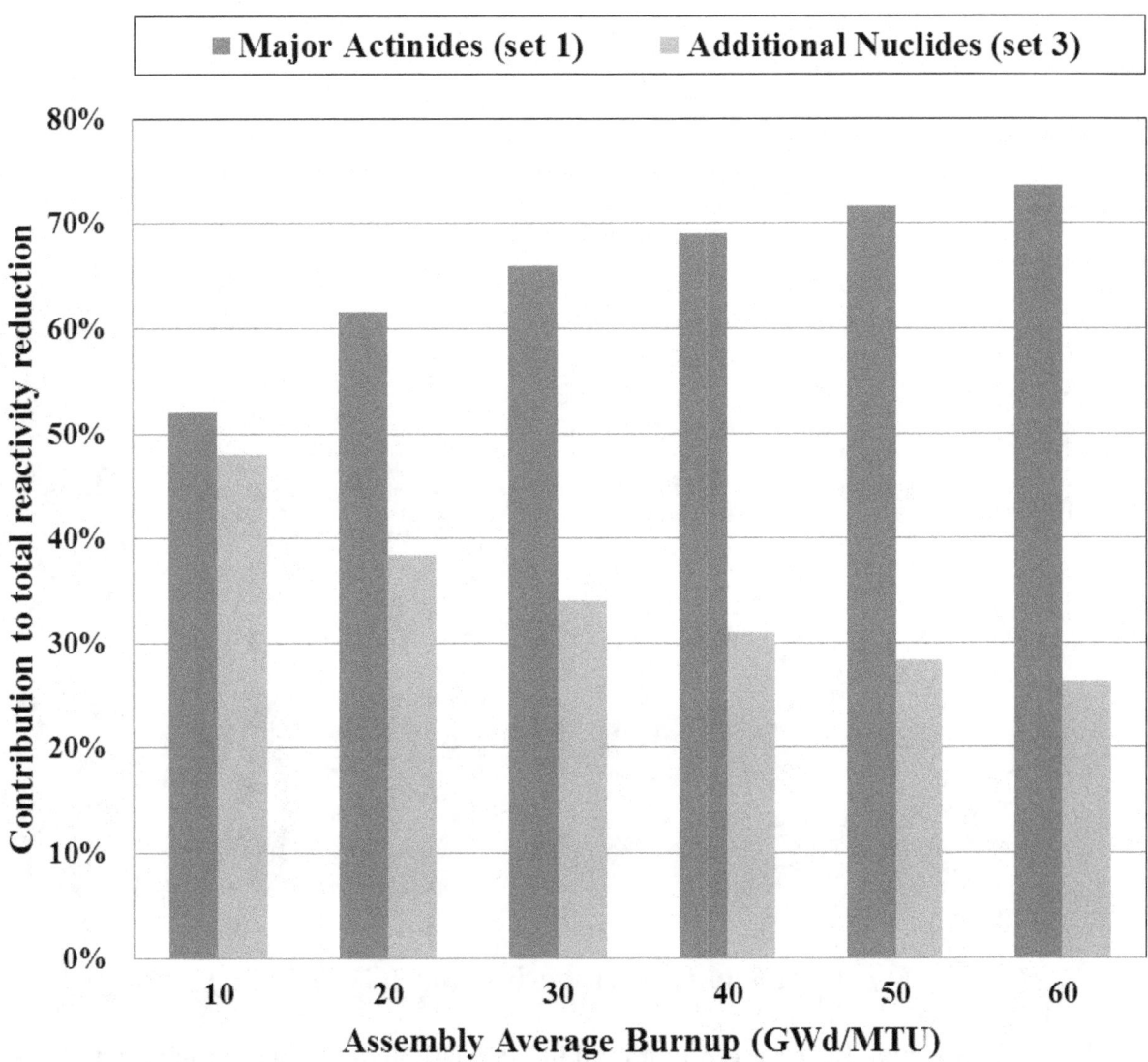

Figure D.12. Individual contributions to the total reduction in k_{eff} for the different nuclide sets as a function of burnup for fuel of 5 wt% ^{235}U initial enrichment with a 5-year cooling time.

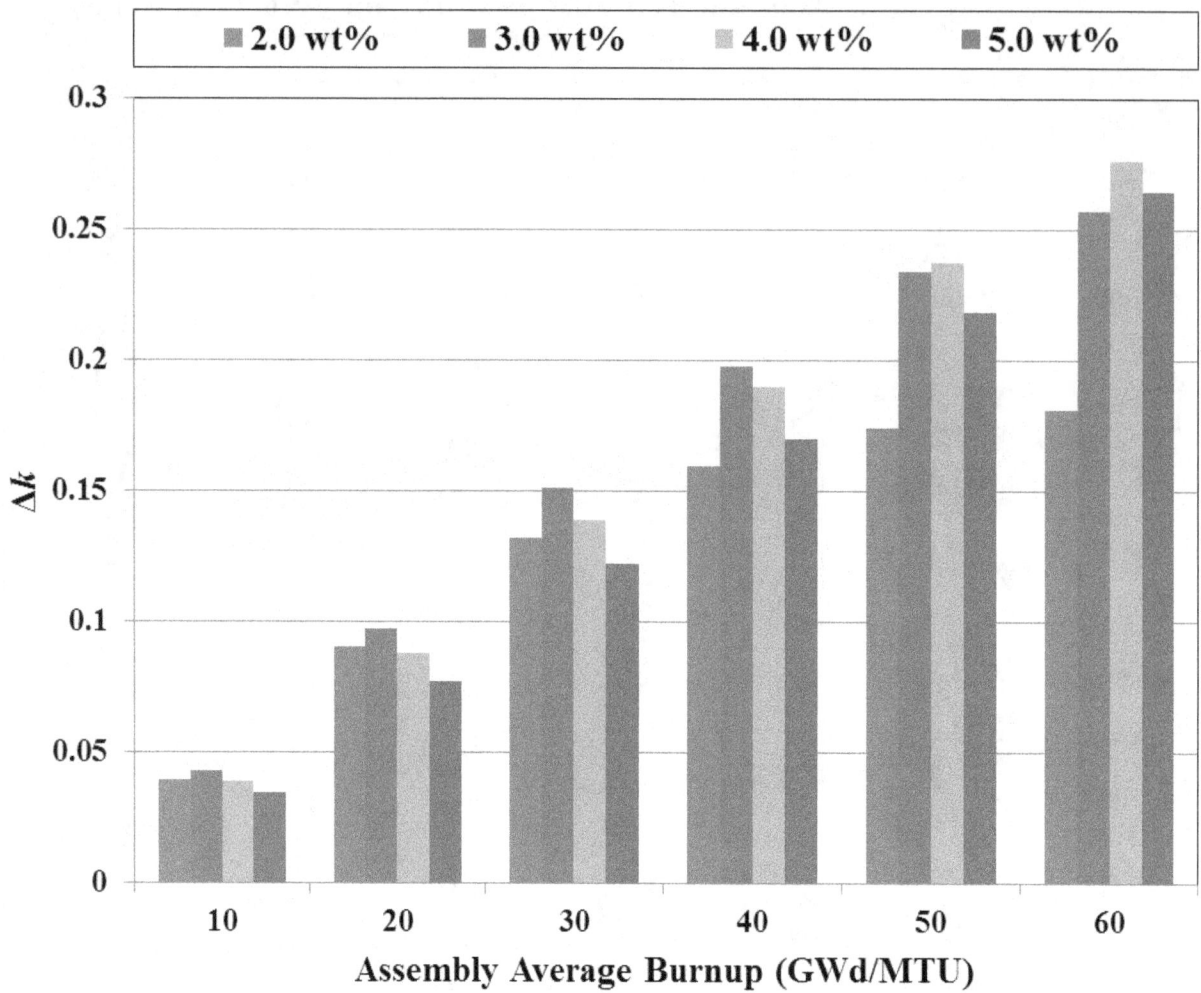

Figure D.13. Δ*k* values (relative to fresh fuel) in the GBC-68 cask due to major actinides (set 1) as a function of burnup for various initial enrichments with a 5-year cooling time.

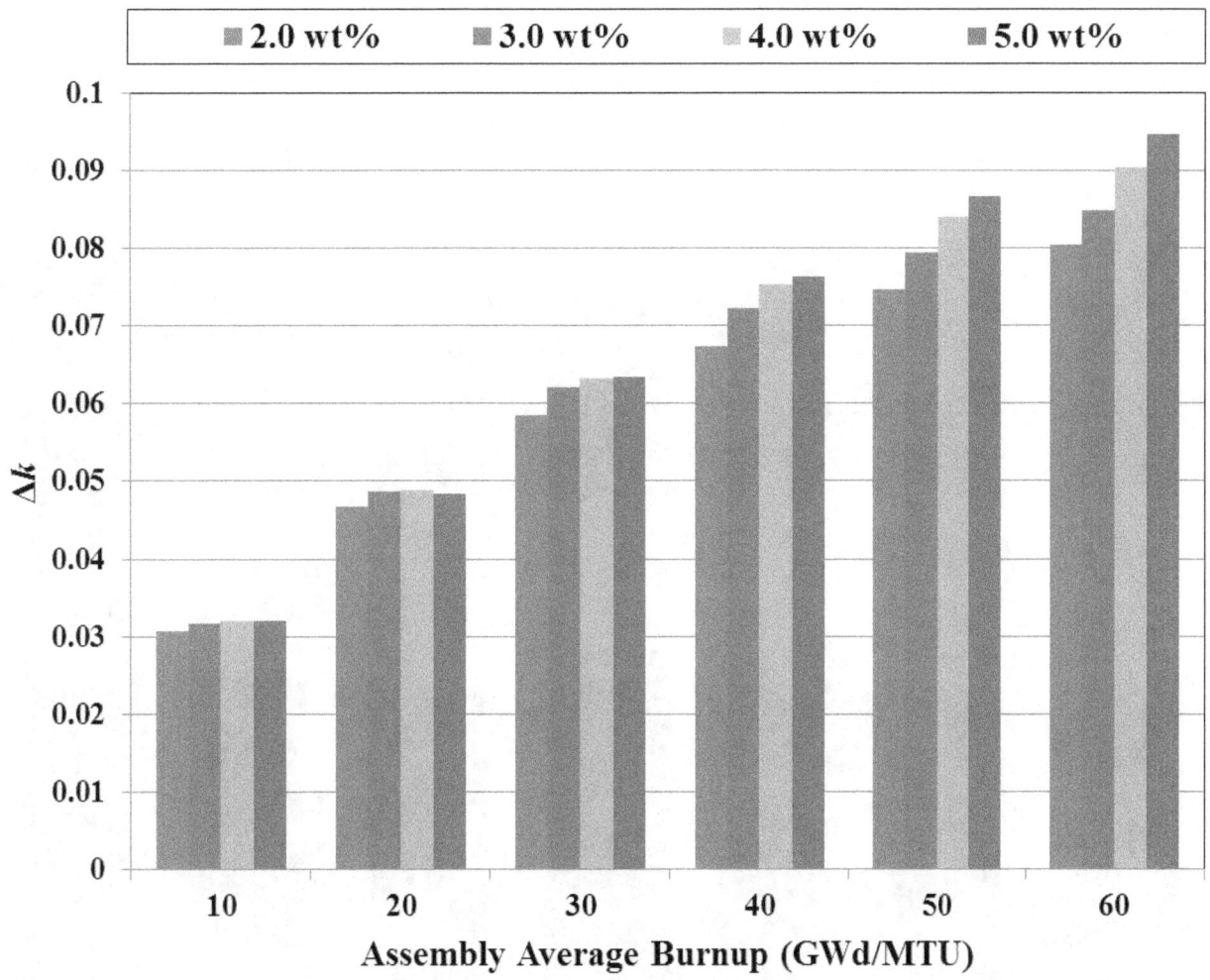

Figure D.14. Δ*k* values (relative to fresh fuel) in the GBC-68 cask due to additional nuclides (set 3) as a function of burnup for various initial enrichments with a 5-year cooling time.

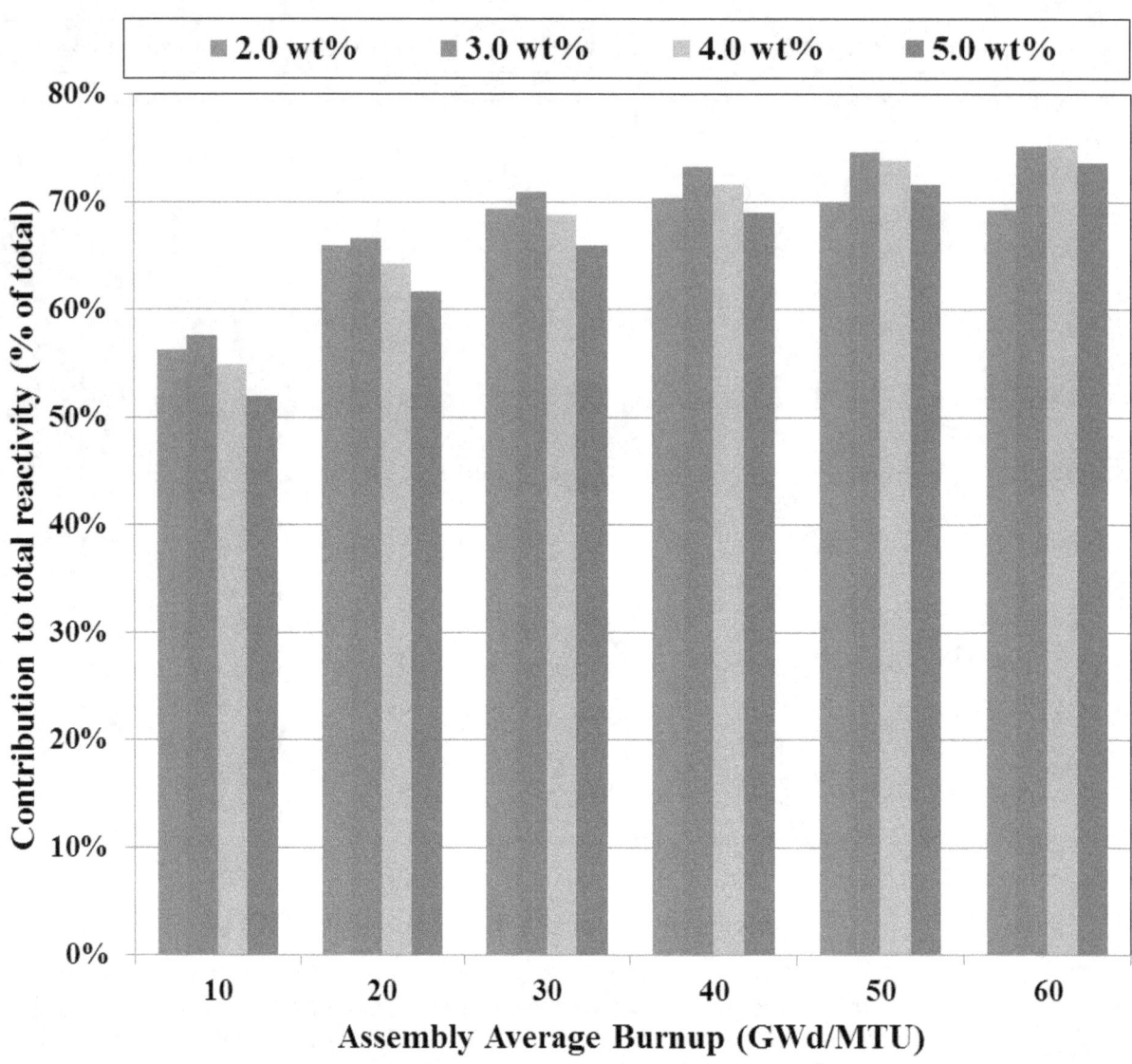

Figure D.15. Contribution to total reduction in k_{eff} due to the major actinides (set 1) as a function of burnup for various enrichments with a 5-year cooling time.

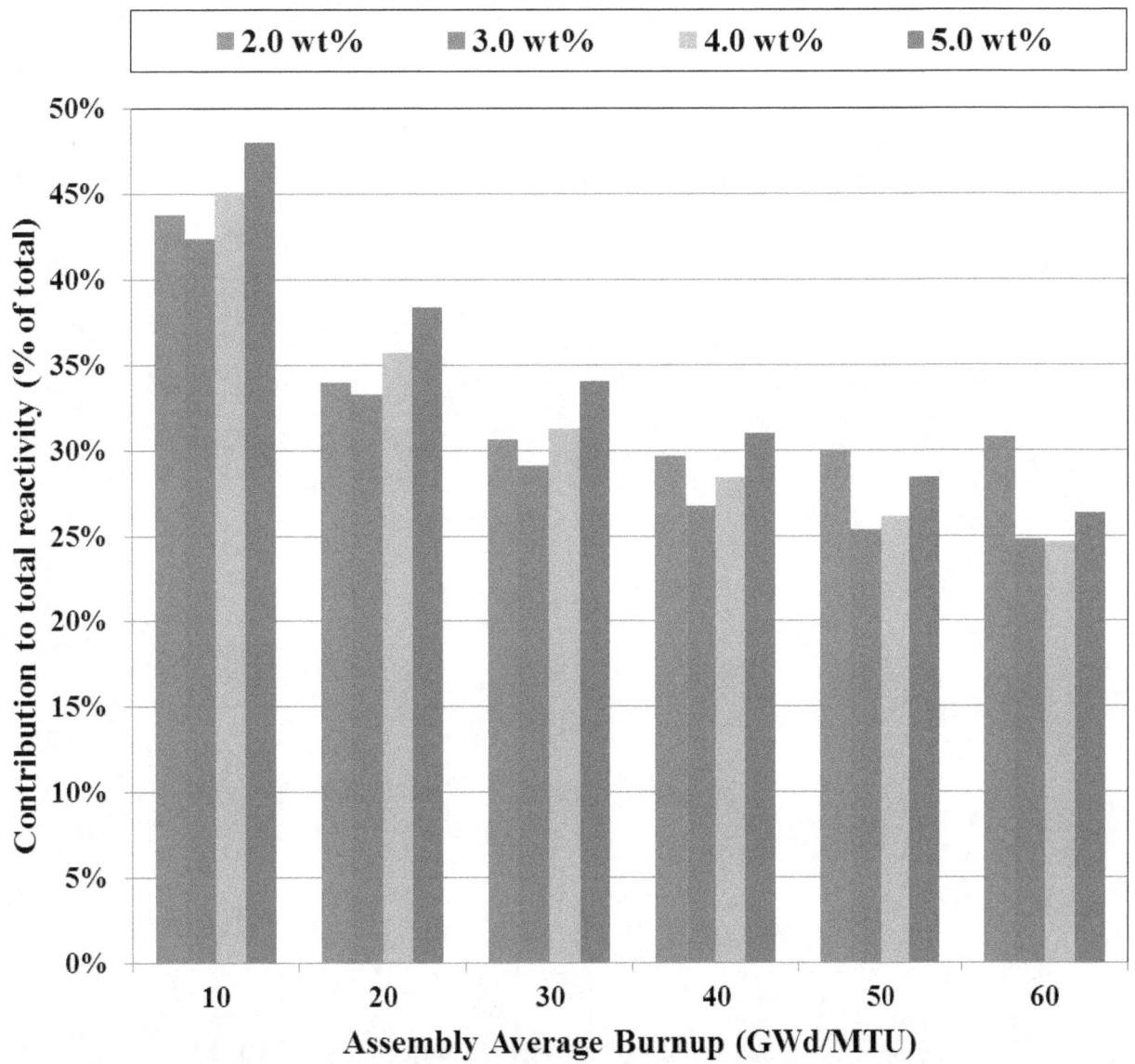

Figure D.16. Contribution of total reduction in k_{eff} due to the additional nuclides (set 3) as a function of burnup for various enrichments with a 5-year cooling time.

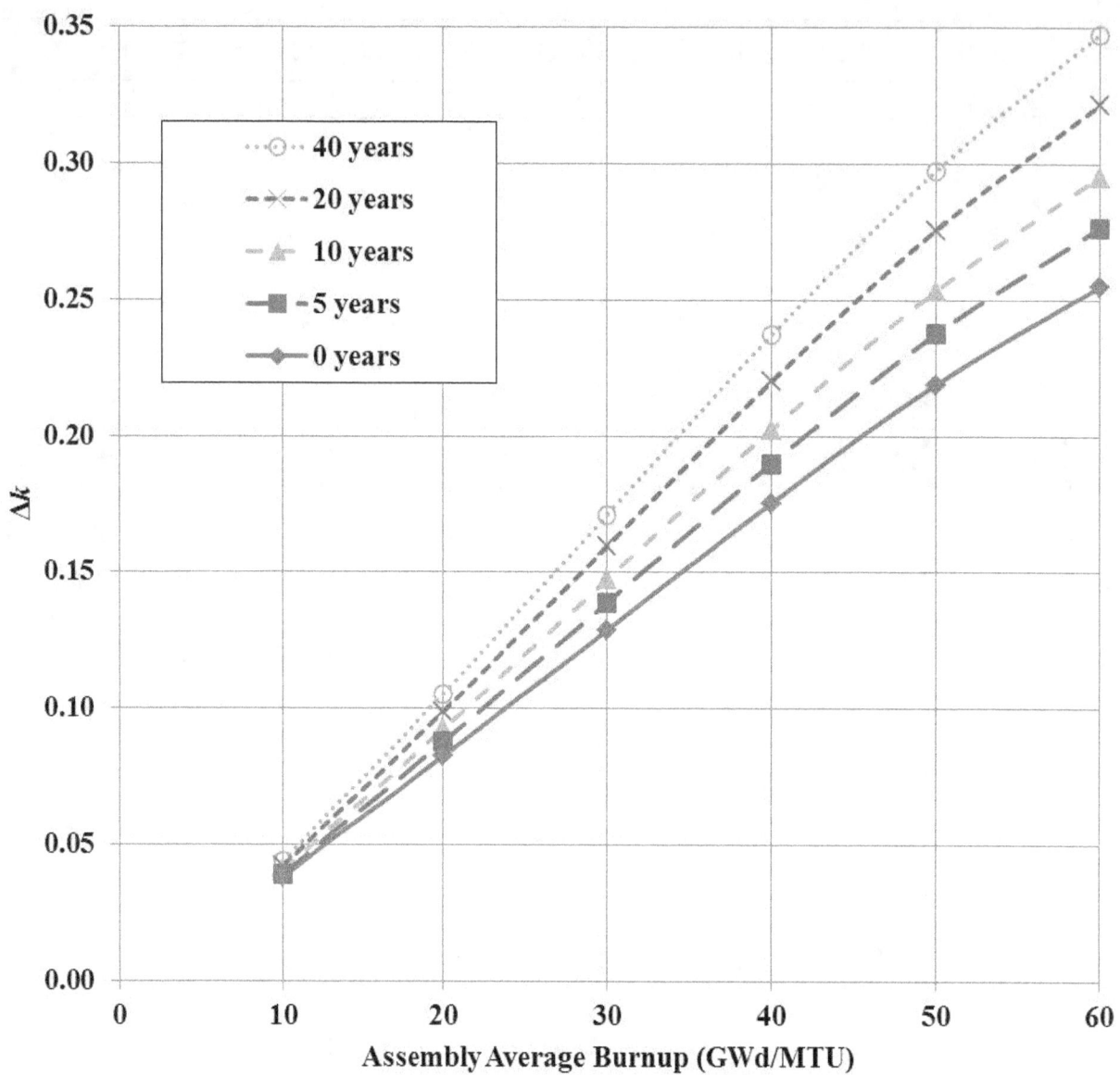

Figure D.17. Δ*k* (relative to fresh fuel) in the GBC-68 cask due to the major actinides (set 1) as a function of burnup for 4 wt% ^{235}U initial enrichment fuel with various cooling times.

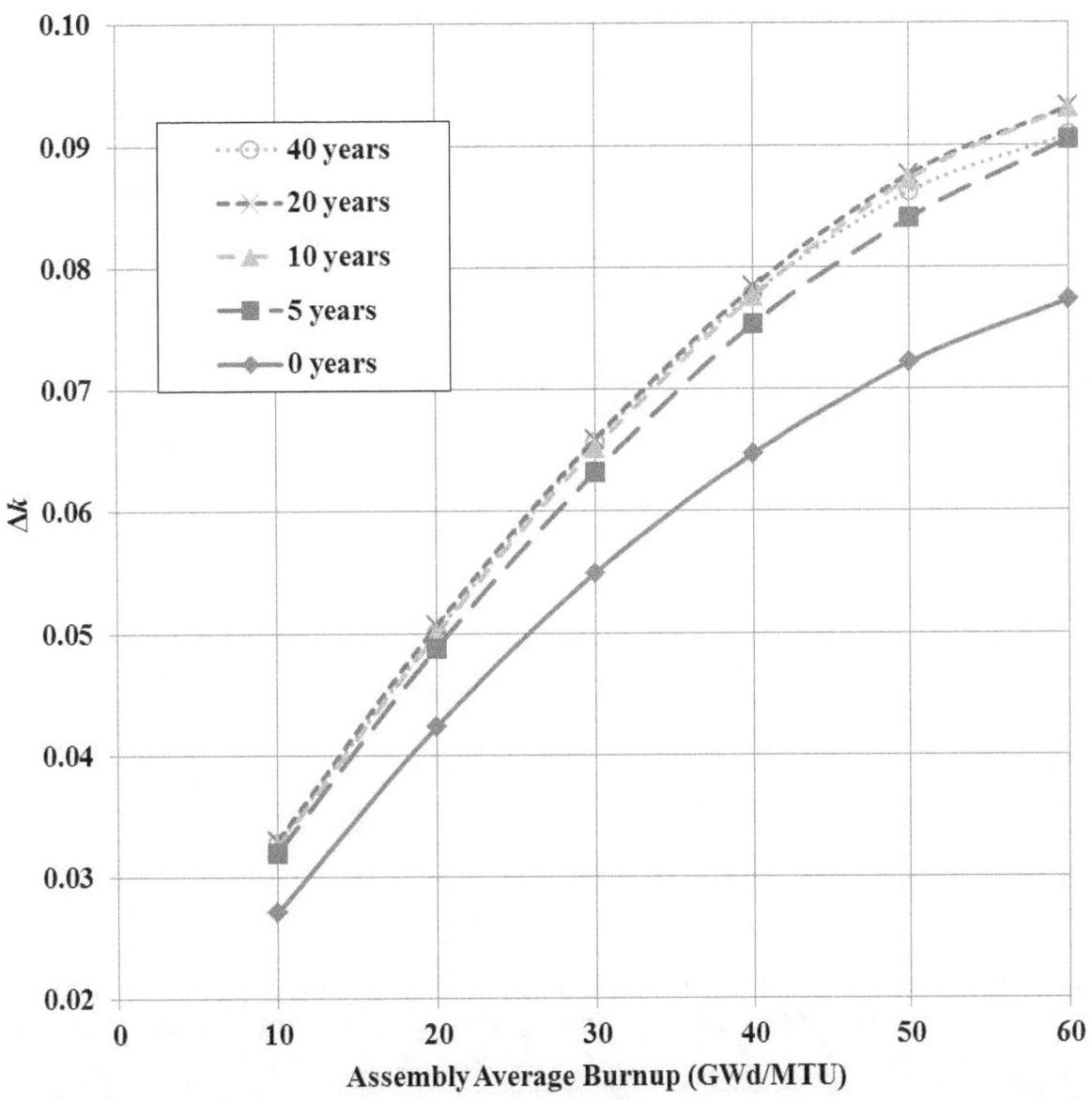

Figure D.18. Δ*k* (relative to fresh fuel) in the GBC-68 cask due to the minor actinides and fission products (set 3) as a function of burnup for 4 wt% ^{235}U initial enrichment fuel with various cooling times.

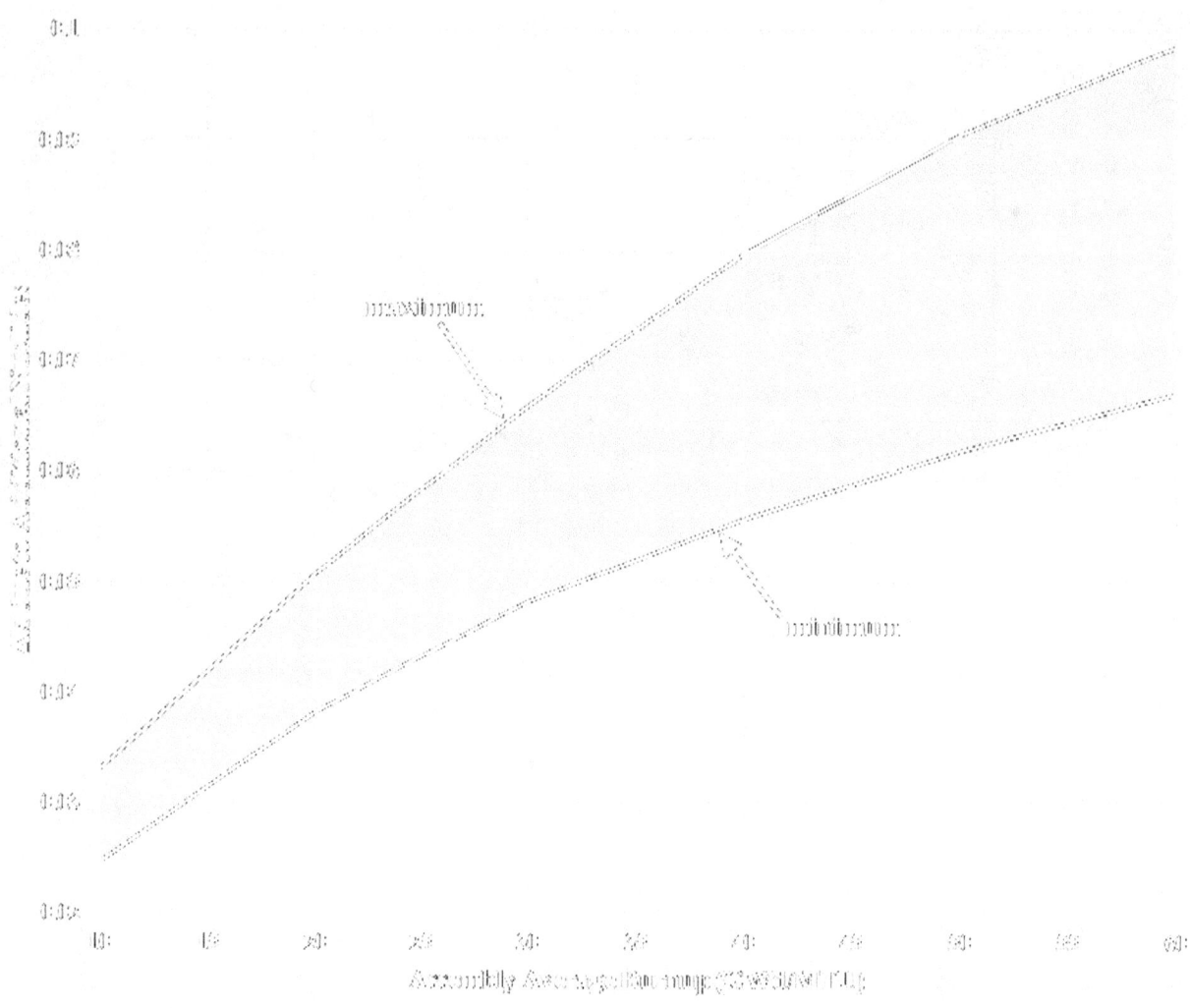

Figure D.19. Range of Δ*k* in the GBC-68 cask due to the additional nuclides (set 3) as a function of burnup for all cooling times and initial enrichments considered.

D.2 APPENDIX D REFERENCES

D.1   RW-859, "Nuclear Fuel Data", Energy Information Administration, Washington, D.C., October 2004.

D. 2 B. L. Broadhead et al., *Investigation of Nuclide Importance to Functional Requirements Related to Transport and Long-Term Storage for LWR Spent Fuel*, ORNL/TM-12742, Lockheed Martin Energy Systems, Inc., Oak Ridge National Laboratory, June 1995.

D.3   T. Suto, S. M. Bowman, and C. V. Parks, "The Reactivity of Nuclide Buildup and Decay During Long-Term Fuel Storage," *Proceedings of the Fifth Annual International Conference on High Level Radioactive Waste Management*, Vol. 2, p. 831, May 22–26, 1994, Las Vegas, NV, (1994).

APPENDIX E. LIMITED NUCLIDE COMPOSITION DATA

Table E.1. Nuclide atom densities (atoms/b-cm) for fuel with initial enrichment of 4 wt% ^{235}U, zero cooling time and various burnups

Nuclide	10 GWd/MTU	20 GWd/MTU	30 GWd/MTU	40 GWd/MTU	50 GWd/MTU	60 GWd/MTU
^{234}U	7.47E-06	6.52E-06	5.63E-06	4.82E-06	4.08E-06	3.43E-06
^{235}U	6.97E-04	4.98E-04	3.43E-04	2.24E-04	1.39E-04	8.22E-05
^{236}U	4.98E-05	8.36E-05	1.08E-04	1.24E-04	1.33E-04	1.36E-04
^{238}U	2.24E-02	2.22E-02	2.21E-02	2.19E-02	2.17E-02	2.15E-02
^{237}Np	1.85E-06	5.00E-06	8.65E-06	1.23E-05	1.57E-05	1.85E-05
^{238}Pu	1.41E-07	7.92E-07	2.16E-06	4.29E-06	6.98E-06	9.89E-06
^{239}Pu	7.75E-05	1.15E-04	1.31E-04	1.37E-04	1.37E-04	1.35E-04
^{240}Pu	1.00E-05	2.59E-05	4.22E-05	5.68E-05	6.85E-05	7.71E-05
^{241}Pu	3.91E-06	1.38E-05	2.36E-05	3.12E-05	3.65E-05	3.97E-05
^{242}Pu	2.60E-07	2.15E-06	6.20E-06	1.21E-05	1.94E-05	2.75E-05
^{241}Am	4.48E-08	3.20E-07	7.64E-07	1.20E-06	1.52E-06	1.69E-06
^{243}Am	1.19E-08	2.24E-07	1.02E-06	2.65E-06	5.13E-06	8.25E-06
^{95}Mo	9.16E-06	2.34E-05	3.67E-05	4.87E-05	5.95E-05	6.93E-05
^{99}Tc	1.51E-05	2.93E-05	4.25E-05	5.46E-05	6.56E-05	7.54E-05
^{101}Ru	1.32E-05	2.63E-05	3.92E-05	5.18E-05	6.41E-05	7.60E-05
^{103}Rh	6.84E-06	1.48E-05	2.19E-05	2.81E-05	3.31E-05	3.70E-05
^{109}Ag	4.98E-07	1.51E-06	2.84E-06	4.38E-06	6.03E-06	7.70E-06
^{133}Cs	1.58E-05	3.09E-05	4.45E-05	5.68E-05	6.75E-05	7.69E-05
^{143}Nd	1.25E-05	2.31E-05	3.11E-05	3.67E-05	3.99E-05	4.12E-05
^{145}Nd	9.27E-06	1.75E-05	2.48E-05	3.12E-05	3.67E-05	4.13E-05
^{147}Sm	4.86E-07	1.59E-06	2.85E-06	3.99E-06	4.88E-06	5.49E-06
^{149}Sm	9.01E-08	9.17E-08	8.80E-08	8.24E-08	7.68E-08	7.24E-08
^{150}Sm	2.86E-06	6.11E-06	9.45E-06	1.28E-05	1.59E-05	1.88E-05
^{151}Sm	3.26E-07	3.76E-07	4.11E-07	4.38E-07	4.62E-07	4.86E-07
^{152}Sm	1.42E-06	2.77E-06	3.76E-06	4.55E-06	5.20E-06	5.76E-06
^{151}Eu	4.98E-10	5.82E-10	5.80E-10	5.52E-10	5.21E-10	5.00E-10
^{153}Eu	7.11E-07	1.96E-06	3.50E-06	5.02E-06	6.39E-06	7.56E-06
^{155}Gd	4.36E-10	1.02E-09	1.86E-09	2.77E-09	3.59E-09	4.32E-09
^{16}O	4.69E-02	4.69E-02	4.69E-02	4.69E-02	4.69E-02	4.69E-02

Table E.2. Nuclide atom densities (atoms/b-cm) for fuel with initial enrichment of 4 wt% ^{235}U, 5-year cooling time and various burnups

Nuclide	10 GWd/MTU	20 GWd/MTU	30 GWd/MTU	40 GWd/MTU	50 GWd/MTU	60 GWd/MTU
^{234}U	7.48E-06	6.55E-06	5.72E-06	4.99E-06	4.37E-06	3.84E-06
^{235}U	6.97E-04	4.98E-04	3.43E-04	2.24E-04	1.39E-04	8.22E-05
^{236}U	4.98E-05	8.36E-05	1.08E-04	1.24E-04	1.33E-04	1.36E-04
^{238}U	2.24E-02	2.22E-02	2.21E-02	2.19E-02	2.17E-02	2.15E-02
^{237}Np	1.94E-06	5.15E-06	8.85E-06	1.26E-05	1.60E-05	1.88E-05
^{238}Pu	1.43E-07	8.22E-07	2.27E-06	4.50E-06	7.28E-06	1.03E-05
^{239}Pu	7.88E-05	1.16E-04	1.33E-04	1.38E-04	1.39E-04	1.37E-04
^{240}Pu	1.00E-05	2.59E-05	4.22E-05	5.69E-05	6.89E-05	7.78E-05
^{241}Pu	3.07E-06	1.08E-05	1.85E-05	2.45E-05	2.86E-05	3.12E-05
^{242}Pu	2.60E-07	2.15E-06	6.20E-06	1.21E-05	1.94E-05	2.75E-05
^{241}Am	8.84E-07	3.28E-06	5.81E-06	7.89E-06	9.33E-06	1.02E-05
^{243}Am	1.20E-08	2.24E-07	1.02E-06	2.65E-06	5.13E-06	8.26E-06
^{95}Mo	1.55E-05	2.96E-05	4.26E-05	5.43E-05	6.49E-05	7.44E-05
^{99}Tc	1.53E-05	2.95E-05	4.27E-05	5.48E-05	6.58E-05	7.56E-05
^{101}Ru	1.32E-05	2.63E-05	3.92E-05	5.18E-05	6.41E-05	7.60E-05
^{103}Rh	8.49E-06	1.67E-05	2.40E-05	3.03E-05	3.55E-05	3.95E-05
^{109}Ag	5.00E-07	1.51E-06	2.85E-06	4.39E-06	6.04E-06	7.71E-06
^{133}Cs	1.63E-05	3.13E-05	4.50E-05	5.72E-05	6.80E-05	7.73E-05
^{143}Nd	1.33E-05	2.39E-05	3.19E-05	3.75E-05	4.07E-05	4.20E-05
^{145}Nd	9.28E-06	1.75E-05	2.48E-05	3.12E-05	3.67E-05	4.13E-05
^{147}Sm	3.63E-06	6.41E-06	8.48E-06	9.92E-06	1.08E-05	1.13E-05
^{149}Sm	1.22E-07	1.28E-07	1.27E-07	1.24E-07	1.20E-07	1.17E-07
^{150}Sm	2.86E-06	6.11E-06	9.45E-06	1.28E-05	1.59E-05	1.88E-05
^{151}Sm	3.20E-07	3.69E-07	4.03E-07	4.30E-07	4.53E-07	4.76E-07
^{152}Sm	1.42E-06	2.77E-06	3.76E-06	4.55E-06	5.20E-06	5.76E-06
^{151}Eu	1.31E-08	1.51E-08	1.64E-08	1.74E-08	1.83E-08	1.92E-08
^{153}Eu	7.22E-07	1.98E-06	3.52E-06	5.05E-06	6.43E-06	7.61E-06
^{155}Gd	2.46E-08	5.98E-08	1.17E-07	1.86E-07	2.54E-07	3.14E-07
^{16}O	4.69E-02	4.69E-02	4.69E-02	4.69E-02	4.69E-02	4.69E-02

Table E.3. Nuclide atom densities (atoms/b-cm) for fuel with initial enrichment of 4 wt% ^{235}U, 10-year cooling time and various burnups

Nuclide	10 GWd/MTU	20 GWd/MTU	30 GWd/MTU	40 GWd/MTU	50 GWd/MTU	60 GWd/MTU
^{234}U	7.48E-06	6.58E-06	5.81E-06	5.17E-06	4.65E-06	4.24E-06
^{235}U	6.97E-04	4.98E-04	3.43E-04	2.24E-04	1.39E-04	8.22E-05
^{236}U	4.99E-05	8.37E-05	1.08E-04	1.24E-04	1.33E-04	1.36E-04
^{238}U	2.24E-02	2.22E-02	2.21E-02	2.19E-02	2.17E-02	2.15E-02
^{237}Np	1.95E-06	5.19E-06	8.91E-06	1.27E-05	1.61E-05	1.89E-05
^{238}Pu	1.37E-07	7.90E-07	2.18E-06	4.33E-06	7.00E-06	9.86E-06
^{239}Pu	7.88E-05	1.16E-04	1.33E-04	1.38E-04	1.39E-04	1.37E-04
^{240}Pu	1.00E-05	2.59E-05	4.22E-05	5.70E-05	6.91E-05	7.84E-05
^{241}Pu	2.41E-06	8.51E-06	1.45E-05	1.92E-05	2.25E-05	2.45E-05
^{242}Pu	2.60E-07	2.15E-06	6.20E-06	1.21E-05	1.94E-05	2.75E-05
^{241}Am	1.54E-06	5.58E-06	9.74E-06	1.31E-05	1.54E-05	1.68E-05
^{243}Am	1.20E-08	2.24E-07	1.02E-06	2.65E-06	5.13E-06	8.25E-06
^{95}Mo	1.55E-05	2.96E-05	4.26E-05	5.43E-05	6.49E-05	7.44E-05
^{99}Tc	1.53E-05	2.95E-05	4.27E-05	5.48E-05	6.58E-05	7.56E-05
^{101}Ru	1.32E-05	2.63E-05	3.92E-05	5.18E-05	6.41E-05	7.60E-05
^{103}Rh	8.49E-06	1.67E-05	2.40E-05	3.03E-05	3.55E-05	3.95E-05
^{109}Ag	5.00E-07	1.51E-06	2.85E-06	4.39E-06	6.04E-06	7.71E-06
^{133}Cs	1.63E-05	3.13E-05	4.50E-05	5.72E-05	6.80E-05	7.73E-05
^{143}Nd	1.33E-05	2.39E-05	3.19E-05	3.75E-05	4.07E-05	4.20E-05
^{145}Nd	9.28E-06	1.75E-05	2.48E-05	3.12E-05	3.67E-05	4.13E-05
^{147}Sm	4.48E-06	7.70E-06	9.98E-06	1.15E-05	1.24E-05	1.28E-05
^{149}Sm	1.22E-07	1.28E-07	1.27E-07	1.24E-07	1.20E-07	1.17E-07
^{150}Sm	2.86E-06	6.11E-06	9.45E-06	1.28E-05	1.59E-05	1.88E-05
^{151}Sm	3.08E-07	3.55E-07	3.88E-07	4.14E-07	4.36E-07	4.58E-07
^{152}Sm	1.42E-06	2.77E-06	3.76E-06	4.55E-06	5.20E-06	5.76E-06
^{151}Eu	2.52E-08	2.90E-08	3.17E-08	3.37E-08	3.54E-08	3.72E-08
^{153}Eu	7.22E-07	1.98E-06	3.52E-06	5.05E-06	6.43E-06	7.61E-06
^{155}Gd	3.63E-08	8.82E-08	1.72E-07	2.74E-07	3.75E-07	4.64E-07
^{16}O	4.69E-02	4.69E-02	4.69E-02	4.69E-02	4.69E-02	4.69E-02

Table E.4. Nuclide atom densities (atoms/b-cm) for fuel with initial enrichment of 4 wt% ^{235}U, 20-year cooling time and various burnups

Nuclide	10 GWd/MTU	20 GWd/MTU	30 GWd/MTU	40 GWd/MTU	50 GWd/MTU	60 GWd/MTU
^{234}U	7.49E-06	6.64E-06	5.98E-06	5.50E-06	5.18E-06	4.99E-06
^{235}U	6.97E-04	4.98E-04	3.43E-04	2.25E-04	1.39E-04	8.22E-05
^{236}U	4.99E-05	8.37E-05	1.08E-04	1.24E-04	1.33E-04	1.37E-04
^{238}U	2.24E-02	2.22E-02	2.21E-02	2.19E-02	2.17E-02	2.15E-02
^{237}Np	1.99E-06	5.30E-06	9.12E-06	1.29E-05	1.64E-05	1.92E-05
^{238}Pu	1.27E-07	7.30E-07	2.02E-06	4.00E-06	6.47E-06	9.11E-06
^{239}Pu	7.88E-05	1.16E-04	1.33E-04	1.38E-04	1.39E-04	1.37E-04
^{240}Pu	1.00E-05	2.59E-05	4.22E-05	5.71E-05	6.95E-05	7.93E-05
^{241}Pu	1.48E-06	5.24E-06	8.94E-06	1.18E-05	1.38E-05	1.51E-05
^{242}Pu	2.60E-07	2.15E-06	6.20E-06	1.21E-05	1.94E-05	2.75E-05
^{241}Am	2.43E-06	8.73E-06	1.51E-05	2.02E-05	2.37E-05	2.59E-05
^{243}Am	1.19E-08	2.24E-07	1.02E-06	2.65E-06	5.13E-06	8.25E-06
^{95}Mo	1.55E-05	2.96E-05	4.26E-05	5.43E-05	6.49E-05	7.44E-05
^{99}Tc	1.53E-05	2.95E-05	4.27E-05	5.48E-05	6.58E-05	7.56E-05
^{101}Ru	1.32E-05	2.63E-05	3.92E-05	5.18E-05	6.41E-05	7.60E-05
^{103}Rh	8.49E-06	1.67E-05	2.40E-05	3.03E-05	3.55E-05	3.95E-05
^{109}Ag	5.00E-07	1.51E-06	2.85E-06	4.39E-06	6.04E-06	7.71E-06
^{133}Cs	1.63E-05	3.13E-05	4.50E-05	5.72E-05	6.80E-05	7.73E-05
^{143}Nd	1.33E-05	2.39E-05	3.19E-05	3.75E-05	4.07E-05	4.20E-05
^{145}Nd	9.28E-06	1.75E-05	2.48E-05	3.12E-05	3.67E-05	4.13E-05
^{147}Sm	4.76E-06	8.13E-06	1.05E-05	1.20E-05	1.29E-05	1.33E-05
^{149}Sm	1.22E-07	1.28E-07	1.27E-07	1.24E-07	1.20E-07	1.17E-07
^{150}Sm	2.86E-06	6.11E-06	9.45E-06	1.28E-05	1.59E-05	1.88E-05
^{151}Sm	2.85E-07	3.29E-07	3.59E-07	3.83E-07	4.04E-07	4.24E-07
^{152}Sm	1.42E-06	2.77E-06	3.76E-06	4.55E-06	5.20E-06	5.76E-06
^{151}Eu	4.80E-08	5.53E-08	6.04E-08	6.43E-08	6.77E-08	7.12E-08
^{153}Eu	7.22E-07	1.98E-06	3.52E-06	5.05E-06	6.43E-06	7.61E-06
^{155}Gd	4.46E-08	1.08E-07	2.12E-07	3.37E-07	4.61E-07	5.71E-07
^{16}O	4.69E-02	4.69E-02	4.69E-02	4.69E-02	4.69E-02	4.69E-02

Table E.5. Nuclide atom densities (atoms/b-cm) for fuel with initial enrichment of 4 wt% ^{235}U, 40-year cooling time and various burnups

Nuclide	10 GWd/MTU	20 GWd/MTU	30 GWd/MTU	40 GWd/MTU	50 GWd/MTU	60 GWd/MTU
^{234}U	7.51E-06	6.75E-06	6.27E-06	6.08E-06	6.13E-06	6.32E-06
^{235}U	6.97E-04	4.98E-04	3.43E-04	2.25E-04	1.40E-04	8.23E-05
^{236}U	4.99E-05	8.37E-05	1.08E-04	1.24E-04	1.34E-04	1.37E-04
^{238}U	2.24E-02	2.22E-02	2.21E-02	2.19E-02	2.17E-02	2.15E-02
^{237}Np	2.08E-06	5.64E-06	9.70E-06	1.37E-05	1.73E-05	2.02E-05
^{238}Pu	1.08E-07	6.24E-07	1.72E-06	3.41E-06	5.53E-06	7.78E-06
^{239}Pu	7.88E-05	1.16E-04	1.33E-04	1.38E-04	1.39E-04	1.37E-04
^{240}Pu	1.00E-05	2.58E-05	4.22E-05	5.72E-05	6.99E-05	8.02E-05
^{241}Pu	5.62E-07	1.99E-06	3.39E-06	4.48E-06	5.24E-06	5.71E-06
^{242}Pu	2.60E-07	2.15E-06	6.20E-06	1.21E-05	1.94E-05	2.75E-05
^{241}Am	3.26E-06	1.17E-05	2.01E-05	2.68E-05	3.14E-05	3.42E-05
^{243}Am	1.19E-08	2.23E-07	1.01E-06	2.64E-06	5.12E-06	8.23E-06
^{95}Mo	1.55E-05	2.96E-05	4.26E-05	5.43E-05	6.49E-05	7.44E-05
^{99}Tc	1.53E-05	2.95E-05	4.27E-05	5.48E-05	6.58E-05	7.56E-05
^{101}Ru	1.32E-05	2.63E-05	3.92E-05	5.18E-05	6.41E-05	7.60E-05
^{103}Rh	8.49E-06	1.67E-05	2.40E-05	3.03E-05	3.55E-05	3.95E-05
^{109}Ag	5.00E-07	1.51E-06	2.85E-06	4.39E-06	6.04E-06	7.71E-06
^{133}Cs	1.63E-05	3.13E-05	4.50E-05	5.72E-05	6.80E-05	7.73E-05
^{143}Nd	1.33E-05	2.39E-05	3.19E-05	3.75E-05	4.07E-05	4.20E-05
^{145}Nd	9.28E-06	1.75E-05	2.48E-05	3.12E-05	3.67E-05	4.13E-05
^{147}Sm	4.78E-06	8.17E-06	1.05E-05	1.21E-05	1.30E-05	1.34E-05
^{149}Sm	1.22E-07	1.28E-07	1.27E-07	1.24E-07	1.20E-07	1.17E-07
^{150}Sm	2.86E-06	6.11E-06	9.45E-06	1.28E-05	1.59E-05	1.88E-05
^{151}Sm	2.45E-07	2.82E-07	3.08E-07	3.28E-07	3.46E-07	3.64E-07
^{152}Sm	1.42E-06	2.77E-06	3.76E-06	4.55E-06	5.20E-06	5.76E-06
^{151}Eu	8.88E-08	1.02E-07	1.12E-07	1.19E-07	1.25E-07	1.32E-07
^{153}Eu	7.22E-07	1.98E-06	3.52E-06	5.05E-06	6.43E-06	7.61E-06
^{155}Gd	4.70E-08	1.14E-07	2.23E-07	3.55E-07	4.86E-07	6.01E-07
^{16}O	4.69E-02	4.69E-02	4.69E-02	4.69E-02	4.69E-02	4.69E-02

NRC FORM 335
(12-2010)
NRCMD 3.7

U.S. NUCLEAR REGULATORY COMMISSION

BIBLIOGRAPHIC DATA SHEET

(See instructions on the reverse)

1. REPORT NUMBER
(Assigned by NRC, Add Vol., Supp., Rev., and Addendum Numbers, if any.)

NUREG/CR-7157
(ORNL/TM-2012/96)

2. TITLE AND SUBTITLE

Computational Benchmark and Estimated Reactivity Margin from Fission Products and Minor Actinides in BWR Burnup Credit

3. DATE REPORT PUBLISHED

MONTH	YEAR
02	2013

4. FIN OR GRANT NUMBER

V6061

5. AUTHOR(S)

Donald E. Mueller
J. M. Scaglione
J. C. Wagner
S. M. Bowman

6. TYPE OF REPORT

Technical

7. PERIOD COVERED *(Inclusive Dates)*

8. PERFORMING ORGANIZATION - NAME AND ADDRESS *(if NRC, provide Division, Office or Region, U.S. Nuclear Regulatory Commission, and mailing address; if contractor, provide name and mailing address.)*

Oak Ridge National Laboratory
P.O Box 2008, MS-6170
Oak Ridge, TN 37831-6170

9. SPONSORING ORGANIZATION - NAME AND ADDRESS *(if NRC, type "Same as above"; if contractor, provide NRC Division, Office or Region, U.S. Nuclear Regulatory Commission, and mailing address.)*

Division of Systems Analysis, Office of Nuclear Regulatory Research
U.S. Nuclear Regulatory Commission
Washington, DC 20555-0001

10. SUPPLEMENTARY NOTES
Mourad Aissa, NRC Project Manager

11. ABSTRACT *(200 words or less)*

This report proposes and documents a computational benchmark model that is used for the estimation of the additional reactivity margin available in spent nuclear fuel (SNF) from fission products and minor actinides in a burnup-credit storage/transport environment, relative to SNF compositions containing only the major actinides. The benchmark problem is a generic burnup-credit cask designed to hold 68 boiling water reactor (BWR) fuel assemblies. The purpose of this computational benchmark is to provide a reference configuration for the estimation of the additional reactivity margin, which is recommended in the U.S. Nuclear Regulatory Commission guidance for partial burnup credit (ISG-8), and document reference estimations of the additional reactivity margin as a function of initial enrichment, burnup, and cooling time. This benchmark model will also be used as the base case in future sensitivity studies exploring the impact of varying reactor depletion parameters and burnup credit analysis modeling approximations. Estimates of additional reactivity margin for this reference configuration may be compared to those of similar burnup-credit casks to provide an indication of the validity of design-specific estimates of fission-product margin. The margin estimates were generated with the SCALE 6.1 package based on ENDF/B-VII.0 nuclear data.

12. KEY WORDS/DESCRIPTORS *(List words or phrases that will assist researchers in locating the report.)*

burnup credit
BWR
fission products
actinides

13. AVAILABILITY STATEMENT

unlimited

14. SECURITY CLASSIFICATION

(This Page)

unclassified

(This Report)

unclassified

15. NUMBER OF PAGES

16. PRICE

NRC FORM 335 (12-2010)

NUREG/CR-7157

Computational Benchmark for Estimated Reactivity Margin from Fission Products and Minor Actinides in BWR Burnup Credit

February 2013

www.ingramcontent.com/pod-product-compliance
Lightning Source LLC
Chambersburg PA
CBHW080303180526
45167CB00006B/2655